Recreational
Gold Prospecting
For Fun & Profit

Gail Butler

D1279964

Gem Guides Book Co. ◆ 1998

To Alvin B. Butler—

my grandfather, a gold prospector

Copyright © 1998, Gem Guides Book Co.
315 Cloverleaf Dr., Suite F
Baldwin Park, CA 91706

Library of Congress Catalog Card # 97-77186
ISBN 0-935182-98-5

Book typography and binding design by Art Waxer, Berkeley, Calif.

Manufactured in the United States of America

Photo Credits
All photos are by the author except as noted below.
Judé Kendrick: pages 78, 79, 80, 82, 92, 165
Marc Davis: front cover, pages 72, 100, 114, 131
Fisher Archives: page 90
Herman Schob: page 93
Mo Hemler: page 64

Contents

Introduction

What comes to mind when you hear the word, *gold?* Perhaps it conjures up images of water-rounded nuggets seen through the clear, sparkling waters of a creek or stream. Or maybe you envision the mellow sheen of lode gold, concealed within the deep places of a mine, lit momentarily by the glow of a miner's lamp.

Although California and Colorado are the best remembered of the western states for their Golden Era chronicles of fortune and misfortune, the Gold Rush in the West and in other parts of the U.S. has never ended—it continues unabated to this day.

This modern-day Gold Rush is one of today's best kept secrets — a secret, however, that is well-known to the few thousands of modern-day gold prospectors who haunt the wooded hills and dry desert washes of countless gold-rich areas looking for that alluring, precious, yellow metal.

Gold is seen by today's prospectors as they swirl aside the layers of black sand in their gold pans. There, gleaming nuggets and flakes reveal themselves to seeking eyes.

While the words "gold prospector" may stir up images of a crusty sourdough miner squatting at rivers edge, pan in hand, or a solitary journeyer, a burro his only companion, in lonely quest for golden riches, today's gold prospectors are seldom crusty and can be almost anyone — including you!

Modern-day prospectors are men, women, and youngsters of diverse life-styles, ages, and backgrounds. They are fascinating people with a sense of adventure. Some are retirees who have always had a hankering to try their hand at panning a little "color." Others are busy professionals who can only get away on weekends, or at most, on an annual vacation. Most recreational prospectors are family groups who enjoy gold prospecting as an inexpensive, fun activity that can be enjoyed by all, regardless of age. Despite the differences of age and life-style, all have one thing in common. They are all recreational gold prospectors.

In years past gold prospecting, the search for gold, was one's path to fame and fortune. Today it is a hobby, an escape into adventure from the hectic day to day struggle. The prospector of today is usually a part-time or "recreational" prospector. Recreational prospecting is gold prospecting for both fun and profit!

This book is your invitation to join the thousands who have discovered that the "Forty-niners" did not get it all. There's still plenty of gold to be found and it's waiting for you!

The Language & Lore
of Gold Prospecting

Gold prospecting, like any other pursuit, has a language all its own. In fact, many phrases in everyday use by non-prospectors come down to us from the Gold Rush era, or from gold mining in general. As an example, we often speak of something being merely "a flash in the pan," or we hope that an idea we're working on "will pan out." Perhaps at the end of the day we may feel.that we've been put "through the mill." Most people use these phrases quite casually, never reflecting on their historical mining origins.

Interesting as that may be, what we want examine in this chapter are terms that are in common use by recreational gold prospectors. These are words and phrases that you should become familiar with. Knowing the definitions of these prospecting and mining terms will allow you to step up from sounding like a mere amateur to conversing like a real "sourdough!"

As you read this list, see how many terms you can find that are part of the common, everyday language used by the non-prospecting crowd. Perhaps you have even used some of them yourself.

Amalgamation: the process by which gold is recovered with the use of mercury. Mercury containing gold is called an "amalgam."

Auriferous: gold-bearing, or containing gold. Auriferous gravels are those which are thought to be rich in gold. Au is the chemical designation for gold.

Bedrock: the country rock "backbone" of an area which may be either exposed to view or covered with sediment.

Bench Deposit: also called "old riverbed," "terrace," or "tertiary" deposits. It is the bank of a river or creek left high and dry when a river changed or deepened its course of flow. It is not something a bird leaves on a park bench!

Bench Placers: same as above.

Black Sand: sand that is made up of heavy minerals such as hematite, magnetite, or other iron compounds. These compounds give the sand its black coloration and are often used by prospectors as an indicator that gold may be present in the area.

Bonanza: a rich find of gold or other valuables.

Color: a prospector often refers to gold as "color." If, while gold panning, you are approached and asked, "Are you gettin' any color?" the question does not refer to the state of your suntan, but if you've gotten any particles of gold. Be forewarned: most savvy prospectors will fib in the negative, especially if they are finding color.

Concentrates: the heavy materials, minerals and gold, which remain behind in the riffles of a gold pan, sluice, dredge, or drywasher. It is these concentrates that the prospector subsequently pans out to recover gold they may contain.

Crevicing: the use of simple tools such as screw drivers, spoons, snifters, stiff wire, and a gold pan to remove gold from exposed crevices and cracks in bedrock.

Dust: fine or flour gold, also referred to as "fines."

Eureka: what a gold prospector is rumored to yell when he finds a bonanza (see above). I believe that this is strictly a legend. No prospector in his or her right mind would inform all within hearing that gold has been found. If a prospector does find gold, he will likely and very quietly inform his companions well out of hearing range of any other prospectors in the area.

Float: rock or mineral that is scattered on top of the ground where the processes of erosion and gravity have left it. It is usually found downhill from its source — usually an outcropping or a vein of the mineral which may still remain intact.

Fool's Gold: often mistaken for gold by beginners because of its brass-like glitter. Usually it is iron pyrite or muscovite mica, which in sunlight can fool the novice.

Glory Hole: an extremely rich concentration of gold or other values.

Gophering: a haphazard and often dangerous type of placer mining in which excavations dug into unconsolidated material in search of gold.

Grubstake: money advanced to a prospector by an interested party to fund a prospecting venture. Today the term most often refers to the job or profession held by a recreational prospector to pay the bills and feed the family so that he, she, or they can afford to go gold prospecting on weekends and vacations.

Hardrock Mining: also called "lode" mining. It is the extraction of gold or other values from solid rock, or veins and pockets in rock.

High-grading: in the past, as well as in mining today, this name was applied to the act of miners stealing gold ore, or other precious materials, from the mines in which they are employed. Today, in recreational gold prospecting, it refers to a partner, or partners, clandestinely removing gold or nuggets from a sluice, or other piece of gold recovery equipment, prior to equal division by all partners in the venture.

Malleable: capable of being hammered without fracturing. This is one of the physical attributes of gold.

Mercury Gold: gold that is covered with mercury as a result of historical mining efforts. It is illegal to use mercury to recover gold from creeks and streams. However, mercury has been used during historical times to recover gold. Prospectors will occasionally find gold that is covered with mercury. Such gold is likely to be found after periods of heavy runoff, which tends to bring up gold that has been deeply buried for many decades.

Nugget: a slug or lump of natural gold that can be heard when it hits the bottom of your vial.

Outcropping: a rock mass that is exposed at the ground surface.

Overburden: boulders, gravel, sand, and silt lying on top of the bedrock of a river, or which cover a gold deposit.

Panning: the art and technique of separating gold from lighter materials using a gold pan.

Pay Dirt: profitable, gold-bearing gravel.

Pay Streak: the rich part of a gold placer, or where gold is concentrated underwater.

Pennyweight: a Troy weight measurement. There are 20 pennyweights in a Troy ounce. Each pennyweight contains 24 grains.

Placer Gold: gold which has eroded free of its host rock. This is the type of gold that the most recreational gold prospectors seek.

Placer Mining: the process of extracting placer gold.

Pocket: a hole, crevice, portion of a vein, or other area where gold has collected in quantity.

Pocket Gold: slightly different meaning from the above. Usually refers to a concentration of "residual placer" gold (see the chapter, "Where to Find Gold, & Why It's Found There").

Prospect: a promising placer area, but one that has yet to reveal its gold values.

Prospecting: the process of searching for gold.

Recreational Gold Prospector: one who prospects for gold for fun and profit.

Retort: in this case, not a fiery verbal comeback, but a process or piece of equipment that removes mercury from gold by use of heat.

Riffles: obstructions or depressions which trap gold. These obstructions and depressions can be naturally occurring, such as those found in bedrock, or can be man-made like those found in gold pans and gold recovery equipment.

Rusty Gold: as its name implies, is gold that has a covering of mineral oxides which may give it a black or reddish coating. This type of gold may be found in desert areas that see little water and where gold has lain undisturbed for a long time.

Sniping: the process of crevicing for gold in underwater crevices.

Sourdough: a term stemming from the Gold Rush days, when most miners and prospectors carried with them sourdough starters for making biscuits, bread, and flapjacks. The pungency of the fermenting starter could be smelled by those many yards downwind of its owner — thus the name "sourdough" in reference to prospectors.

Specific Gravity: a term which refers to the relative weight and mass of a mineral sample as compared to an equal volume of water.

Stringer: refers to a thin vein of ore material.

Tailings: the buildup of rock and gravel which accumulates at the tail end of a sluice, dredge or drywasher. Also the mound or pile or waste rock and debris left at the tunnel of a mine after ore has been removed.

Tommyknockers: from the lore of Welsh miners, tommyknockers are the elemental beings, or gnomes, which are said to inhabit mines, caves and other holes in the earth. Tommyknockers were thought to be responsible for cave-ins, blow outs and other mining disasters. They were also believed to be able to prevent these disasters to those who won their favor. They were traditionally appeased by miners who left tins of chewing tobacco. It is said that nonsmokers and non-chewers may propitiate the tommyknockers by leaving licorice sticks. "Red vines" licorice also works well for appeasement purposes.

Troy Ounce: the weight of gold and other precious metals is measured in Troy ounces, or fractions of Troy ounces, such as grains and pennyweights. As a comparison, 12 Troy ounces are equal to a pound instead of the standard 16 Avoirdupois ounces that we generally use when measuring ordinary commodities.

Unconsolidated Materials: as opposed to solid rock. Usually a collection of rock, dirt, and gravel which may occur as loose sediment or lightly cemented material.

Vein: a lengthy occurrence of ore in host rock.

Of course there are more terms, but these are the ones most generally used. Don't be put off by the number of terms either. Just study them, and their definitions, and you will be surprised how easily they will spring to mind and roll off your tongue when you start conversing with other prospectors. Be careful though…frequent use of these terms will get you the undivided attention of any nearby prospectors!

Where to Find Gold, & Why It's Found There

"Gold is where you find it," or so the old saying goes. But gold is not always found where you may be looking! In order to narrow your search to those areas most likely to contain gold, it pays to know something about the nature, or properties, of gold and how it deposits and concentrates along a water course. Gold does not deposit evenly along a river or stream, but is found concentrated in certain types of places. A successful prospector must become familiar with the types of places where gold tends to concentrate and why it is found there.

A knowledgeable prospector is able to match what he or she knows about the properties of gold to an ability to "read" the topography of a location known to be gold-bearing. To properly read an area one must have a working knowledge of the properties of gold and how those properties cause gold to move and deposit within the natural environment.

First of all, gold is heavy! It has a specific gravity of between 15 and 19, which varies slightly depending on its purity, and upon the specific gravity of the other metals with which it is alloyed. As a comparison, pure lead, which is quite heavy, has a specific gravity of about 11. To understand specific gravity, it is useful to know that minerals are assigned various "properties" which are used as the criteria for identifying them. One of these properties, specific gravity, is a measurement of density. Specific gravity is measured as an element's "weight by unit volume" and this is how it works. For example, we weigh one cubic centimeter of water. We find that this small volume of water weighs one gram. Any mineral's specific gravity can then be figured by weighing one cubic centimeter of mineral and comparing that to the same volume of water.

So, if a cubic centimeter of an element, say hematite (an ore of iron), weighs 5.26 times as much as an equal volume of water, the specific gravity would be 5.26. Now lets take our sample of gold, which is alloyed with some silver and weigh it against an equal volume of water. We find that our gold sample weighs 17.6 times more than the water. Therefore, the specific gravity of our gold sample would be 17.6. A cubic centimeter of pure, unalloyed gold would weigh 19.3 times as much as the same volume of water. That's it! That's all there is to specific gravity.

Don't worry; you won't have to measure specific gravity to be a recreational gold prospector, but it is useful to understand what the term means. And basically what it means is that gold is really heavy, far heavier than the heaviest things that you can expect to see in your gold pan. Below is a brief list of some of the items – after you've panned off the detritus of rocks, gravel and dirt — that are likely to be found in your gold pan and their approximate specific gravity.

Item Specific Gravity

Pyrite (fools gold) 5.01
Hematite (iron oxide) 5.26
Magnetite (iron oxide) 5.20
Galena (lead ore) 7.57
Garnet 3.61 to 4.05 depending on species
Lead (bullets, fishing weights, etc.) 11
Gold 15.6 to 19.3 depending on purity

You will see that gold is by far the heaviest (in specific gravity) item that is likely to be found in your gold pan! It is not actual weight or size that determines how an object settles and deposits among other material in your gold pan, but specific gravity! This makes gold nearly impossible to lose once it is in your gold pan or gold recovery equipment.

How does gold get to the places where it is found by recreational gold prospectors? Certain natural earth processes, such as water movement, freezing and thawing, gravity, wind, earthquakes, etc., erode gold from its host vein. Due to these influences the vein material degrades and decomposes, freeing gold. Gold, despite its softness and malleability, is quite indestructible. Gold does not rust, fracture, or dissolve (except in strong acids). Unlike silver or copper, it does not oxidize. Let's see how earth processes, or weathering, create placer gold from lode gold.

Rain falls, entering minute cracks and fissures in the exposed vein material, then freezes, widening those fissures. Over time gold is freed from the source vein because of the differences in adhesion and cohesion between gold and the vein material. Gravity may affect gold by pulling it gradually, inexorably, down a slope. This may be accelerated by earthquakes (especially in California). Wind may abrade exposed vein material, gradually wearing it away with tiny, airborne dust particles.

Rain falls again and flows downslope, washing the freed gold down along with mud and gravel. This gold-laden runoff flows to low spots where it may eventually enter a river or creek and, if the current is strong, may be moved quite a way downstream. If the current is slow, however, gold will deposit close to where the forces of erosion, water runoff and gravity, left it.

When gold enters a strong current, it bumbles and bounces along the river bottom until something alters or slows the current enough for gold's specific gravity to cause it to fall from the current and settle onto the gravel of the river bottom. Once settled, gold does not rest long on the river gravel because the riverbed is in constant motion.

The sand and gravel that form the sediments of the riverbed are in a perpetual state of agitation caused by the friction of water flowing over it. When a river is in flood, rocks and boulders may be caught and carried by the raging torrent. The banging together of boulders will jar the river bottom even more, causing gold to settle more rapidly through the sediments after it has fallen from the current. This agitation of the uppermost layers of the riverbed actually keeps them in a state of suspension, allowing gold to rapidly settle ever deeper. In most cases gold will sink out of sight within seconds of settling upon the riverbed.

Over a period of months and years, depending on the depth of the river sediments, gold will work its way through the layers of sand, gravel and rock, to the very bedrock of a watercourse. Here it eventually comes to rest and may settle into cracks, fissures, and crevices, or other low spots, that occur in the bedrock.

Remember that the process of gold settling to bedrock may be constantly interrupted. In seasons of heavy rainfall, or maximum runoff from snow melt, huge volumes of water will course through a river channel, churning and redepositing the sediments. When this happens, gold too may be swept up, temporarily being redeposited nearer the surface before it begins the process of redepositing. Also, new gold will be brought down from the slopes and deposited near the surface.

It is for this reason that recreational gold prospectors turn out in greater-than-usual numbers to prospect after periods of heavy rainfall, flooding, or runoff from snow melt. These periods, right after massive water movement, herald a bonanza for prospectors, for gold is found in quantities and concentrations far above the average.

On the other hand, during a period of drought spanning several years or longer, gold will not be redeposited near the surface but will continue to work its way ever deeper, putting it farther out of the reach of the gold prospector.

An important point to keep in mind is that anywhere that water flow is altered or slowed by objects or topography, gold will settle out. Therefore, the places that tend to accumulate gold are the inside bends of waterways, the downstream sides of large boulders, rocks, and tree stumps, and amongst the roots of trees and shrubs. Gold will often be found embedded in the meshed roots of grasses, secreted in the cracks and crevices of exposed bedrock, or deposited in sandbars that are built up as the result of the confluence of two rivers or creeks. Sandbars may also be caused by a diverging mid-river current; even the moss that clings to rocks also traps and holds gold. All of these places are excellent spots to begin your search for gold.

Let's return to a key point: rivers, creeks, and streams will run with varying volumes of water according to seasonal factors. All watercourses have a tendency at some point to overflow their banks when

Gold is found along the bends of water courses, where the flow of water slows.

flooded; in midsummer, some may reduce their flow to a narrow channel. In areas such as southern California, summer heat and lack of rain will mean that some creeks run only during the rainy season and into early summer, becoming completely dry through summer and fall. This is an excellent time to prospect for possible mid-creek paystreaks.

Because watercourses tend to shrink within their channels or overflow their banks, depending on seasonal factors, gold may be found at varying distances from the present water's edge. A prospector must be aware of this and plan the search accordingly. This is really what "reading the river" is all about – the ability to visualize how the river may run under a variety of conditions caused by seasonal changes in water volume and velocity. There are clues and tools which help the prospector to accomplish an accurate river reading. The first tool is the topographic map.

Examination of a topographic map gives the prospector a "bird's eye view" of the watercourse. The twists, turns, constrictions, and widenings of its channel may be readily seen by consulting a map.

Constriction of a watercourse means that water will flow faster, thus carrying gold farther. Where the course widens, water flow expands and correspondingly slows. In this case, gold falls from the flow far sooner, beginning its settling process. A "topo" map will also reveal rapids and waterfalls – places where gold may flow quickly, then settle within calm pools at the base of rapid movement. The topo map will tell you if the flow is constant or intermittent. It will indicate "contour intervals." The contour interval of a map is the difference in elevation between adjacent contour lines. This information is useful when attempting to figure the most likely spot for gold concentration to occur. A useful fact to remember is that a gradient — the degree of slant of a hillside — of 30 feet to the mile tends to be richest in placer gold concentration. Thus, at a glance, the prospector may familiarize himself with a length of river, or the surrounding topography of a placer, and begin to predict quite accurately the most likely spot to find gold!

Once at this spot, the prospector then needs to take a good look around. If it is midsummer, observe whether the river has been reduced to flowing narrowly, leaving many places dry that were formerly covered with moving water. Ask yourself how high the river may have been during peak runoff in early spring. Do you see a flood plain – a boulder strewn area where the river overflowed its banks at one time, or perhaps seasonally over a period of time?

Try to imagine what the river flow looked like during each of its seasons, and where various obstacles might have reduced or altered the current during each season. The boulder that is high and dry during summer may have been underwater in spring; beneath it may be a repository of nice gold nuggets. Try to discover the highest point of water flow, compared to where the watercourse flows presently.

A clue to look for in figuring the high water mark is to look at trees and shrubs along the bank. Is detritus caught among the limbs? Are branches, logs and other debris strewn about upon the banks? Do you see a relatively flat, dry expanse of boulders? Their presence indicates a flood plain — a spot where a river at flood stage flowed over its banks.

That which is out of place, uprooted, or strewn about in unlikely places will possibly indicate where water has run. These are all great places to begin your search for gold.

What effect do objects, which alter the flow of water, have on gold deposition? A log or boulder, or the roots of a tree that grows at water's edge, each has the ability to effect the speed at which water flows by slowing it, or creating eddies and backwashes where water contacts these objects. When flowing around a large boulder, for example, water hits the upstream side, and speeds up slightly as it eddies around the rock. Some particles of gold may drop out and settle upon the sediments against the upstream side of the rock because of the slight eddying of the current when it first encounters the rock. The water then speeds up again as it moves around the boulder, slowing as it comes to the downstream side of the boulder. This results in the majority of the gold that had been caught in the current dropping out on the lee-side of the boulder. A similar process occurs with other obstacles encountered by flowing water.

The topography of the watercourse also alters the flow of water, causing gold to settle out and concentrate. Water is slowed at the lower end of an inside bend, usually just past the center mark, so gold may be concentrated there. As the water enters the next bend gold will be deposited along the first half of the inside curve of that one. Water is slowed by sandbars, thereby depositing gold, as well as sand and gravel that contribute to the growth of the sandbar.

Paystreaks — placers of gold that settle out of the current mid-river — are prospected for and recovered with dredges or underwater gold detectors. Here gold drops out of the current due to obstructions, such as boulders, exposed projections of bedrock, or slight elevation changes in the riverbed that act to alter water flow. These may either slow the flow or create eddies, both of which have a similar effect on gold. The rule of a 30-foot-per-mile gradient applies profitably to prospectors who seek these mid-river concentration of gold! For prospecting all other gold placers, the prospector will use a gold pan, sluice box, rocker, highbanker or metal detector.

Whether you prospect mid-river paystreaks or streamside placers, you must keep in mind that although your placer may be quite rich, the underlying bedrock may be even richer in gold. Bedrock — the solid mass of rock that underlies most water placers — is not really solid. It is riddled with cracks, cavities, and fissures which may become packed with gold. Gold deposited in bedrock may not be churned up and redeposited during seasonal flooding, but may lie

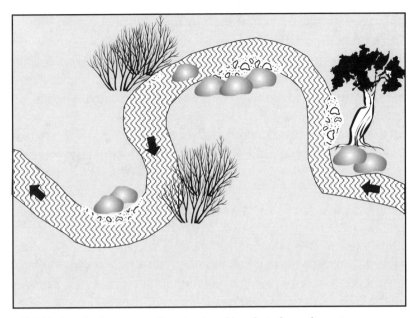

Gold is found where water flow is altered by obstacles or by watercourse contours.

Gold is found near the confluence of two streams.

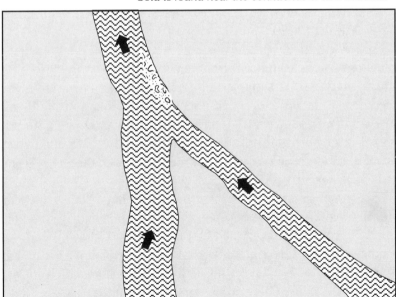

undisturbed for years, trapped and packed tightly within the crevice. When you prospect a placer, you may dig down a few feet and hit bedrock. Don't stop there. It is likely that you will find even more gold secreted in the cracks, then in the placers themselves.

As a prospector, you may eventually encounter what is called "false bedrock." This is usually a layer of hard or cemented composite material overlying the actual bedrock. Dig through this and you will likely find another layer of gold-bearing gravel. A hardened clay made of calcium carbonate, called "caliche," is quite commonly found in the desert areas and separates various layers of gold-bearing gravel.

Also, when opening fissures in bedrock, it is likely that you may encounter a malleable, clay-like substance at the bottom. Dig up this clay and pulverize it completely in your gold pan. Then pan it along with whatever other material is in your pan. This clay must be pulverized and rubbed to destroy its mass, as it has the rumored ability to "steal" gold from your pan. The belief that clay steals gold from your pan is a common notion among prospectors. I am not sure if clay can actually do this, since I have always taken precautions to insure that it does not.

Is it possible that clay can somehow alter the specific gravity of gold with its own, lesser, specific gravity? Most prospectors accept as fact that this indeed does happen and have not tested the truth of it any more than I have. We all simply take the precaution of breaking and rubbing the clay in the water of our gold pans, until it loses its cohesion. This is a quick process in that clay is usually found in minute quantities. Caliche, which is found in greater quantities in some desert placers, can be ground up, then panned out to recover the gold it may contain. The clay found at the bottom of bedrock crevices is actually pulverized silica. There are a variety of nifty ways to find and extract gold from bedrock crevices, and these will be covered in the next chapter.

It has been said that where the prospector finds black sand, there too will be found gold. While finding black sand in your gold pan is a good sign that gold may indeed be present, it in itself is not a sure indicator that gold is nearby. Black sand is made up of heavier metallic elements such as iron, and will usually be found in a gold pan, or in the riffles of gold recovery equipment, along with gold. This is due to the high specific gravity of the metallic elements that compose black sand. Gold is often found in veins in which other metallic elements occur. When gold erodes out of a vein, it is often accompanied by these other metallic elements, as they often tend to occur together in nature. Keep in mind that iron is a very common mineral and that while it is often occurs in veins with gold, it more often occurs without gold. The components of black sand are most often oxidized magnetite and hematite, hence its black color.

I have prospected and recovered gold in areas where no black sand occurred; only fine white quartz sand obscured the gold in my pan. This is rare, for in most areas that I prospect copious black sand must be swirled aside to reveal the heavier gold particles and nuggets lying beneath. Telltale deposits of white quartz sand are called "white runs," and are definitely worth investigating, should they occur in an area you are prospecting. If you are prospecting in South America, or the southeastern U.S., you may also run across heavy yellow sand, which is composed of monazite, (cerium-lanthanum-thorium phosphate). In huge deposits, these minerals are commercially mined in other parts of the world.

In areas of intermittent flow, such as desert washes, black sand may lie on top of the ground. It is a good indication of where to prospect due to the high specific gravity of both black sand and gold, as both tend to settle out in the same types of places and under the same conditions. Visible deposits of black sand may form a swirl-shaped pattern, and are good clues to where water slowed and may have also deposited gold during runoff from a storm or flash flood. Generally, in areas of constant water flow, black sand will be well mixed with surrounding sand and gravel and be found only when you begin to "process" or pan the concentrates of gold recovery equipment.

Prospecting desert washes is not that much different from prospecting rivers, creeks, or streams, except that any water you use, you must bring. This of course necessitates the use of a different type of "waterless" recovery equipment.

Desert washes indicate where rain water, hopefully containing gold, drained off of nearby hills and mountains and flowed out upon the desert. Reading desert washes is very similar to reading rivers. Note the contours of the wash, the gradient, obstacles which may have altered the flow, and if black sand is visible (it may not always

"Sample panning" is the best way to find the highest gold concentrations within an area.

be). These are just some of the clues that will be used to prospect for gold in the desert. Another clue is outcroppings, masses of country rock which are exposed at the surface. Lodes within outcroppings may have eroded, leaving residual or alluvial placers to be prospected for gold values.

In a nutshell, water moves gold, whether it flows continuously or intermittently. By developing the skill of "reading" water flow — how water volume, velocity, and natural objects and contours, affect that flow — the prospector becomes extraordinarily adept at finding where gold concentrates and deposits. This is the skill that determines whether your gold prospecting will be both fun and profitable! Believe me, there is a connection between the two factors, although they are not totally dependent upon each other. It is possible to have a fun prospecting trip and not find one flake of gold. In this case, one enjoys what IS available — the company of family members, friends, the beauties of natural surroundings, cozying up around a warm campfire at night with companions. Also, food cooked over an open campfire has an exquisite flavor that cannot be duplicated anywhere else! There is much to enjoy and revel in when out in nature. Finding gold is only the cherry on top of an all-around enjoyable, family or group-oriented activity. I have found that beauty and fun are best enjoyed when shared with others.

Don't get me wrong. Solitary prospecting can be a time to repair from work stresses, to be alone with your thoughts, to solve personal problems and spend time alone with nature. When I first began recreational gold prospecting, I was the only one I knew who did it. Gradually, I was able to interest some of my friends in going. Also, I met like-minded people in the field, some of whom have become lasting friends. Children take readily to gold prospecting, making it the ideal family activity.

Now, I rarely prospect alone, unless I feel like "getting away from it all." In some cases, my friends who all have the typical "9-5-Monday-through-Friday" grubstake, cannot accompany me on midweek trips. So, I who am fortunate enough to be a full-time prospector, if I choose, will pack my little four-wheel drive Suzuki Samurai and head out on my own with few regrets.

Until now, we have been discussing placer gold — gold that has eroded out of its source vein. As a gold prospector you will need to know the difference between "lode" and "placer" gold. Lode gold is that which is still contained within its source rock, often a quartz vein "contact" within a "country rock" matrix. A "contact" is when another rock material is extruded or injected, via earth processes, into the fissures of an already existing rock type. When the surrounding country rock erodes, you will see the contact as a band of dissimilar material running through the country rock. The contact will vary in coloration and substance. "Country rock" is a term that is often used to describe the existing rock-type that underlies the land. The term "host rock" is another that is used to convey the same idea. The term "contact zone" will describe the area where the two materials meet.

Many placer prospectors — most recreational gold prospectors are placer prospectors — occasionally extend their search to include lode gold. All placer gold — at one time, lode gold — may, under the proper conditions, be traced back to the original vein — a vein that may still contain gold, provided that it has not completed eroded. There are several techniques for tracing gold to its source which will be examined in the next chapter.

If the original vein no longer exists, placer gold — that is, found along a watercourse — may have been deposited and redeposited over the course of centuries, or even millennia. The topography of areas changes considerably over time, and deeply buried lodes may eventually surface. The earth processes are ongoing, and, although unlikely, it is not impossible that the recreational prospector may discover a new source of lode gold.

Let's forego a complex geology lesson that would go beyond the scope required for successful recreational gold prospecting. We will, however, cover some basic "placer geology" that will serve the recreational gold prospector in his or her quest for gold.

There are several types of placers that recreational prospectors should be aware of, although there is no need to force yourself to remember their names. Just be aware of the types of placers and where they are likely to be found:

Residual Placers are weathered from their source, and either remain at the site of weathering, or may be carried down slope by gravity and running water. These are usually deposited elsewhere at a distance from their original source. Residual placers are also the source of "pocket gold." The technique for prospecting and recovering pocket gold will be covered in a later chapter.

Alluvial Placers are similar to residual placers but remain upon the hillside from which they eroded, also a source of the yet-to-be-explained pocket gold.

Bench Placers (also referred to as Bench Deposits) are formed from water placers. These were laid down by running water. Over a period of time the current altered its course, possibly carving a new one elsewhere, or deepened its original course of flow, leaving gold trapped high and dry in ancient bank sediments high above the present-day flow.

Flood placers are those which were deposited by a watercourse overflowing its banks and depositing gold a distance away.

Desert Placers are those that are deposited and moved in arid regions due to intermittent water movement, such as flash flooding from occasional cloudbursts.

Beach Placers are gold placers which are deposited and concentrated by wave action, such as the gold deposits in the beach sands of Nome, Alaska.

Glacial Placers are those created by the melting of glaciers. Glacier deposits may contain gold that is far removed from its original source, carried by the glacier and then deposited when it melted. These placers are not usually known for a high concentration of gold values, as glacial movement has a tendency to scatter gold, rather than concentrating it. Glaciers often move gold so far from its point of origin that it is impossible to locate the source vein.

Water Placers are the most common and diverse of the various types of placers and the ones which should form the backbone of study for most recreational prospectors. Study with extra care the section earlier in this chapter on "reading the river."

A better understanding of where placers are found than what they are called forms the essential starting point for successful gold prospecting.

What is "fool's gold" and how does one identify it? Fool's gold is that which may deceive the novice into thinking that he or she has found gold. The deceit is accomplished because of a golden coloration that is somewhat reminiscent of gold. Real gold has a warm, heart-melting, golden hue. It is rich in depth and warmth of color. It maintains its golden warmth whether in direct sunlight, in shade, or under the illumination of a dimly-lit camp lantern.

Fool's gold is usually brassy in comparison to the true, rich golden tones of real gold. Pyrite is iron sulfide and crystallizes in cubes. You may find these cubes in your gold pan due to their specific gravity of 5.01. Pyrite oxidizes to brown. Gold neither oxidizes or rusts. You may also find small flakes of pyrite which consist of either shattered cubes or pyrite which failed to crystallize. Pyrite is always a pale, brassy color. It is harder than gold and more brittle. The edge of a knife blade pressed onto gold will dent or flatten it due to gold's property of malleability (gold is the only malleable yellow metal). The same, done to pyrite, will shatter it. Also, pyrite will tend to lose its slight gold coloration, fading to a muddy hue, when observed in shade out of direct sunlight.

Another property to test for is "streak." When dragged across the unglazed back of a piece of tile, a pyrite fragment leaves a brown streak, a telltale sign of its iron content. Gold leaves a golden-yellow streak on unglazed tile.

While gold does crystallize, it is unlikely that you will ever find gold crystals in the placer fields, due to its malleability, which allows it to be too easily mashed into the lumps and slugs that we refer to as nuggets. Rarely, crystalline gold is found in undisturbed lode veins and is highly prized by collectors for its specimen value, a value that far exceeds the value of the gold itself. When gold crystallizes it forms octahedral and dodecahedral (pyramid-shaped) forms. It may distort into dendritic (fernlike) shapes, or form long, slender "wires."

Muscovite mica (hydrous potassium aluminum silicate) is another mineral that is sometimes confused with gold. Mica is not metallic, but sometimes glints with a yellowish or brassy hue in sunlight. Placed in the shade, mica fragments fade to black and seem to disappear. Mica's luster may sometime appear to be metallic, but with closer examination this proves untrue, for this luster is actually more "greasy" or glassy than metallic. Mica forms in thin laminated layers that are somewhat flexible and tough. It is one of the common components of granite and will shatter when pressed with the edge of a knife blade. Mica also has a very light specific gravity of about 2.7 to 3.0. Therefore, it will pan out with all the lighter materials in your gold pan and be gone long before you get to black sand and gold. People are fooled when they see mica lying beneath the water of a shallow stream where sunlight highlights its brassy color. Mica leaves a white streak that is all but invisible on unglazed, white tile.

Other differences manifest when you drop these three substances into a water-filled vial. Gold plunges immediately to the bottom of the vial. If the piece is large enough, such as a tiny nugget, you may hear an audible "click" as it hits the glass bottom of the vial. A little chunk of pyrite drifts more slowly down and is generally silent when it hits bottom. A flake, or chunk of mica leisurely floats down, taking quite a bit longer to drift and twirl its way down. Gold, on the other hand, descends so fast that it is nearly impossible to see while falling.

So now you know why gold is found where it is — and you know how to tell the difference between fool's gold and the real thing. Now it is time to learn how to sample an area for gold and how to recover it, all of which means — it's time to GO PROSPECTING!

Basic Techniques of Gold Prospecting

Now that we understand where gold is found and why, it's time to learn to use a gold pan. Once this is accomplished, the world of recreational gold prospecting is wide open to you. Where you go from here is up to you. There are a variety of methods used to recover gold. All you have to do is pick the type, or types, of methods that most appeal to you. But before we get to that, you must learn to use a gold pan.

The gold pan is both the beginning and the end of the gold prospecting process. The main function of the gold pan is to sample for gold concentrations for subsequent recovery by other types of equipment. The second use is for the cleanup of the concentrates that are collected by sluices, dredges, drywashers, and highbankers. A gold pan, in most cases, is not a gold-recovery tool. Notice that I said, "in most cases." There are a couple of exceptions that we will get to later.

There are many sizes, shapes, materials, and colors of gold pans on the market today. This can make choosing one confusing. However, when armed with a few facts, the prospector will easily be able to choose a pan that will be serviceable and comfortable to use.

One of the most important things to do is to select a gold pan that is the proper size. A mistake made by both pro and novice is to purchase a pan that is too large. Using one that is too big will result in sore back, shoulder, and arm muscles which is likely to translate into a growing distaste with the panning process. Gold panning should not be a painful or discouraging experience. It should be, and is, a fun, exciting, and often rewarding venture.

A general rule used in picking a gold pan for a woman is that she is usually most comfortable using a pan that is 10 or 12 inches in diameter. A youngster is best able to handle a pan that is 10 inches, or less, in diameter. A man, or husky teen, should select a 12-inch pan. Pans that are larger than 12 inches are cumbersome and awkward to use for any length of time. It may not feel cumbersome in the store when it is empty, but think of wielding the pan hour after hour, loaded with sand, water, and gravel and you will begin to get the idea.

Selecting the proper size of gold pan.

A simple way to choose the right size pan is to pick up the one you are considering for purchase and hold it so that the open side of the pan lies against your forearm. Rest the rim against your elbow crease and note where the opposing rim lies. Ideally, the pan should fit between the elbow and wrist crease. If it lies between your wrist crease and mid-palm, you have probably found a pan that will be comfortable to work with. If the rim extends past the base of your fingers, the pan is too large and you should select the next smaller size.

There are square and rectangular-shaped gold pans, but it is the traditional round pan that is favored by most gold prospectors. A square pan can be used in all applications of prospecting. A triangular pan is only useful for sampling for gold concentrations. The triangular pan comes nearly to a point at its base and will tip over if you try to empty a bucket of concentrates into it for subsequent panning.

Gold pans are manufactured from both metal and plastic. Prospectors are divided on which material is their favorite. Usually a prospector will favor the type with which he or she learned to pan. Metal pans rust and are not suitable for prospecting with a metal detector. They also need to be "blacked," or burned, in order to darken the light-colored steel sufficiently to allow the prospector to easily see gold against its surface. Plastic pans are sometimes too light in weight. A stiff gust of wind can easily topple one into a river where it may be lost downstream. These are the basic pros and cons of metal and plastic gold pans. Let's examine them in more detail.

Metal Gold Pans

These are my favorites. In all applications of prospecting, with the exception of nugget shooting (using a metal detector to find gold), I use metal pans. Again, this is purely a personal preference.

Metal pans are available in copper and steel. Steel pans are inexpensive, usually under five dollars. Copper pans are used *only* for gold amalgamation—recovering gold using mercury — a process that will be covered in detail later. Also, copper pans are about four times as expensive as steel, and are so soft that rocks can scratch through their outer surface revealing a line of bright, fresh copper. This, glimpsed briefly through black sand, can appear deceptively like a stringer of gold flakes.

A new steel pan will be covered with a coating of machine oil to protect it from rusting until purchased. This oil must be removed before use or it will not allow gold to behave properly in the pan. The slightest bit of oil remaining in your pan may not allow gold to break the surface tension of water in the pan during the panning process. This results in gold flakes being panned off rather than sinking to the bottom of the pan for later retrieval.

When you get your pan home, wipe off the oil using a paper towel. Set the pan on the back burner of a gas stove and turn up the heat to medium-high. This process of cooking the pan is called "blacking." Let the pan black, or cook, until it turns a fairly uniform blue-black color. At this point, turn off the flame and let the pan cool on the back burner. Do not attempt to rapidly cool the pan by immersing it in water as it could possibly warp.

You may wish to repeat the blacking process several more times until the pan is sufficiently dark in color. Should any oil remaining on the pan begin to smoke, turn on the vent fan and open the windows, or let the pan cool, then wipe it again to remove more of the oil before repeating the blacking process.

What I prefer to do is remove all the machine oil by first wiping the pan with a paper towel, then washing it in hot soapy water to remove the last vestiges of oily residue. Then I lightly coat the pan with either cooking oil or lard, again wiping up any excess, so that only the thinnest coating of oil remains. Next I black the pan. This way no possibly harmful fumes from the machine oil are emitted into the atmosphere of my home during the blacking process.

Blacking may also be done on a camp stove. The old timers used to black their pans in the ashes over the hot coals of the campfire. I tried this once and ended up with a sufficiently blacked, but also oily, ash-coated pan. It is much easier and cleaner to do this on a stove.

If you have children in the house, black your pan when they are at school. Be sure to let anyone who is in the house know that you are blacking a gold pan, and under no conditions are they to touch it. A painful burn would result if someone unknowingly picked up a hot pan.

Blacking the pan also helps to prevent rust. Notice that I said "helps." Your pan will still rust, but not so quickly or completely as it would without blacking. If you notice a rust spot on your pan, it means that at some point it was put away wet. In this case, rub away the rust with a paper towel or disposable piece of cloth. It is not usually necessary to use steel wool unless the pan is in an advanced state of corrosion. Apply a thin coat of oil, rubbing up the excess with your paper towel, and re-black the pan. The best and simplest way to prevent rust is to thoroughly dry the pan before putting it away, or just let it sit in the sun to air-dry before packing it up.

With time and use your steel pan will mellow to a rich, dark-brown patina. It will actually look like the gold pans that the 49ers used!

Plastic Gold Pans

The main advantage to plastic is that it doesn't rust. Still, a plastic pan needs a little prepping before you take it out the first time. Plastic pans too come with a thin coating of oil, from the molding process, which must be removed. You may notice that water will have a tendency to bead up on the surface of a new plastic pan. This beading up is undesirable and can be eliminated by merely scrubbing the inside of the pan with a light abrasive cleanser such as the type that is used to clean sinks and bathtubs. That's it. Nothing more is required to prep a plastic pan!

Plastic pans can be put away wet, retrieved months later, and be none the worse for wear. In fact, I recommend plastic pans for the "occasional" prospector or fisherman who packs along a gold pan just to satisfy his curiosity or pass the time while waiting for fish to bite. These pans are also good for kids, due to the near-indestructibility of plastic. Kids may not be as careful as adults about putting their pans away dry. So for the kids, plastic is usually preferred.

Plastic gold pans come in a variety of colors. For most prospecting applications, the best colors are dark-green and black. These two colors are easiest on the eye that is scanning intently for gold, which is easily seen against these two colors.

Gold pans made of plastic are the only type you can use when nugget shooting, as the detector will target a metal pan rather than the nugget you may be trying to isolate. Another advantage to plastic pans is that they do not conduct cold from frigid water as readily as metal.

About the only drawback to plastic pans is that some are too light in weight. A gust of wind may send one tumbling into the river. This is easily countered however, by anchoring it with a rock. Sometimes while holding a plastic pan up to my face to scan for gold, gusts of wind will make it difficult to hold steady, and thus focus my eyes on the task at hand. Metal pans are heavy enough to forestall this minor wind movement.

Plastic pans will melt, so keep them well away from the campfire. Also, some of the cheaper models, made of softer plastic, will bow out of shape when filled with soil and gravel. Eventually the cheaper plastic pans become permanently warped.

Drop-Bottom Gold Pans

I heartily recommend that whether you purchase steel or plastic pans, you make sure that the pan you select is a "drop-bottom" style. Both plastic and metal pans come with drop-bottoms and these are the best for retaining gold. The bottoms of conventional gold pans are straight and even where they meet the sides of the pan. The bottom section of drop-bottom pans "pooches" out a little and extends downward slightly, about a quarter-inch past a clean, straight intersection with the sides of the pan. The Garrett Gravity Trap plastic gold pan has a drop bottom, as do the plastic and metal pans manufactured by Keene Engineering. The drop-bottoms of these pans make nice nugget traps and retain fine gold exceptionally well.

Many plastic pans, and a few metal ones, have ridges called "riffles" molded into the sloping sides of the pan. These riffles make panning away fine gold less likely, but they also slow the panning process considerably. A drop-bottom pan enhances gold retention, but does not slow the panning process.

The faster you can pan, the sooner you will be able to process the gold-bearing concentrates of your gold recovery equipment, or finish your sample panning. This is an advantage when you consider the prospectors' age-old axiom, "The more dirt you move, the more gold you get." Allow me to also add, "The sooner you move the dirt, the sooner you will get the gold!"

The Art of Panning for Gold

You now know something about the various types of gold pans on the market, their advantages and minor disadvantages, and how to pick the size right for you. It's time to learn how to use one!

Gold panning is an art that is very simple to learn. It forms the foundation of most aspects of recreational gold prospecting. Gold panning is merely a combination of agitation to help gold settle, and gentle sluicing movements to slide unwanted material from the pan.

You will discover that gold prospectors use a variety of methods of agitation and sluicing that are uniquely their own. I will give you the basic method for gold panning. From this you are likely to develop a technique that is yours and yours alone.

For your first gold panning session you will need your gold pan, a predetermined number of metal shavings (lead shavings are great due to their high specific gravity), or BBs, some dirt, and a water-filled receptacle, such as a washtub. Be sure that your washtub is large enough in diameter, and deep enough, to allow for comfortable and free movement of your gold pan. Fill your pan about a third full of soil (a small shovel full of dirt from the garden will do nicely), metal shavings, and whatever rocks happen to be in the shovel full of soil.

Once you become more accomplished at panning you may wish to increase the amount of material you put into your pan, but for now you just want practice panning and retaining your practice pieces of metal, rather than seeing how much dirt you can process. Also, garden dirt will be messy to pan because of the organic material. Gravel and sand near streams and rivers has been washed of much of its organic material and dust, which otherwise will float on top of the water or muddy up the pan.

Slowly sink your filled pan into the washtub. You may let it rest on the bottom, kneading the contents with both hands, or hold it sub-

Fill the gold pan about half full of gravel and dirt (hopefully, gold-bearing!).

merged with one hand while kneading and working the contents with the other. At this point dust and organic material will rise to the top of your wash tub and obscure what you are doing. Just ignore it and keep going.

Knead the material until all is completely wet, being careful not to push anything out of the pan. Once the material is thoroughly wetted and separated, raise the pan and tip it, emptying out just enough water so that the contents don't slosh over the side once you begin to agitate or circulate the pan.

Holding your pan level, circulate it in a counterclockwise motion without sloshing out any of the contents. As the contents start to swirl, the process of gold (or metal) settling to the bottom of the pan begins. You will notice that rocks and gravel inside the pan will begin to rise to the surface and ride upon the swirling mass.

Some prospectors use a rapid, wrist-rocking motion, which accomplishes the same purpose as the swirling motion. This second method is more likely to result in the pan's contents slopping over the rim. Try

Knead the contents of the pan to break up any clods of dirt.

both and see which one you prefer. Both movements create the agitation which causes gold to begin to settle to the bottom of the pan.

Now start to shake the pan gently sideways, tilting it slightly, to help settle and slide the contents into the lowest area (according to the angle at which you are now holding it) of the gold pan. The gold at

Add water and swirl the pan in a circular motion.

this point is held in the crease where the side of the pan intersects with the bottom, or in the drop-bottom portion. Holding the pan at a shallow angle, you will submerge it so that the water in your washtub will be able to sluice or wash off the top layers of material.

Begin to move the pan forward and backward gently, keeping it steady at a shallow angle beneath the surface of the water. It is this gentle forward and backward movement that will allow the water in your washtub to remove, or "sluice" off, the top layers of material.

Gentle motion forward and back "sluices" material from pan gradually. Gold stays in the riffles. The pan is held at a slight angle.

As the bulk of the material begins to creep toward the rim of your pan, bring the pan level and agitate the contents, or repeat the counterclockwise swirling motion, to again resettle gold into the bottom/side crease of the gold pan. You may wish to replace the counterclockwise motion with a sideways shaking motion once the material in your pan begins to diminish — usually about half way through the panning process.

Simply repeat the swirling, or agitation, and sluicing motions, until only black sand remains in the pan. Any stubborn gravel may be picked out with the fingers. These panning motions are repeated and modified in intensity until only black sand and gold (or metal shavings) remains.

To see metal shavings or gold beneath black sand, have just enough water remaining in the pan so that you can swirl, or wash, the sand easily over the bottom of your gold pan. With a semicircular movement, wash the sands across the bottom of your pan, spreading them out. Spreading the black sand this way will show you if gold is present; as you swirl aside the black sand you should see your metal pieces, or gold, trailing behind the sand as it flows to the opposite side of your pan.

When you can pan, and retain the same number of metal shavings, or BBs, that you put in, you have mastered the art of panning. In fact, due to gold's relatively high specific gravity, it will be easier to keep gold in your pan than the metal shavings. You may feel a bit awkward with your panning skill, but you will become faster and more confident with practice. You will naturally alter your panning techniques, as required by the amount and type of material that you have placed into your gold pan. You will adjust your panning technique to accommodate shallower, or deeper, water, slower or faster current. Always look for a relatively still area out of the main current to pan, as a strong current will wash the contents from your pan and leave you with nothing.

Be assured that, although you may still feel awkward with your new-found panning skills, you have already become proficient enough to go out into the field and prospect for gold.

Keep in mind that there may not be as much black sand in your garden soil, if any, as in most gold-bearing areas. The main thing is to retain the bits of metal that you threw into the pan to begin with. As we discussed in the previous chapter, not all gold-bearing areas have black sand. In some areas it is possible to pan right down to gold!

To briefly recap:

The swirling, counterclockwise motion is used to bring unwanted rock and gravel to the surface of the gold pan for subsequent panning off. It also begins the process of settling gold to the bottom of the pan.

Slight tilting of the pan and a gentle sideways shaking motion assist gold in settling into the crease at the bottom, and behind any riffles that may be molded into the sides of the pan. Settling is further assisted by tipping the pan slightly back towards level while shaking it to bring the contents away from the rim.

A gentle sluicing motion reduces the volume of unwanted materials in your pan. These motions are repeated until the bulk of material has been reduced enough that you can scan for gold.

There you have it! That is all there is to the art of gold panning.

The next thing to learn is how to easily remove gold from the pan. Many pieces of gold are too small to be picked out with fingers. These tiny pieces, particularly if they are difficult to see without a loupe, are called "fines." The really small stuff that looks like a yellow powder is called "flour" gold. Flour gold requires a different recovery process, usually using mercury. Many recreational prospectors don't even bother recovering the small amount of flour gold. Flour gold is usually seen when ore is crushed, then panned out.

For the fines, or small flakes, you will need a thin, tapered artist's brush. If necessary, trim the bristles to a point and break, or cut, the handle so that it is only about four inches long from handle end to brush tip. The perfect brush for picking flakes of gold from a gold pan is the type that women use for putting on liquid eye liner. Guys, talk to your girlfriends about that one!

The method used by most prospectors to remove gold from the pan is to rest the pan comfortably, and well balanced, on your lap. Hold a water-filled, clear glass vial over the pan. Be sure the vial cap is removed, but close by. With the other hand holding your tiny brush, dab up a flake of gold and place it into the vial. As soon as the water in the vial contacts the bristles of your brush, the gold flake will fall rapidly to the bottom of the vial. Watch to see that this occurs. If not, swirl your brush in the vial a bit and the gold will then fall off. You can pick up several flakes at once with your brush and deposit all into your vial together. Fine-tipped tweezers can be used to pick up bigger flakes and tiny nuggets. Of course, fingers are great for picking up most nuggets.

What do you do if you get out into the field and you've forgotten to pack the little brush that you use to remove gold flakes from your pan? Merely dry a finger, usually an index finger, that you will use to pick up the gold. Just wipe your finger across your pants to dry it, if wet. Then, press it down on top of the flake of gold that you wish to remove from your pan. Raise your finger and check visually if the gold is sticking to it. Position your finger, and the gold upon it, tightly over the opening of your vial. Tip the vial so that the water in it contacts the part of your finger where the gold is, then tip the vial back upright and watch your flake of gold descend to the bottom.

Check to make sure that the gold has gone into the vial, then dry your finger again and pick up the next piece, and so on. Gold will stick to any dry object — in this case, your finger. The fact that gold tends to stick to dry surfaces is another reason why you must thoroughly wet the contents of your gold pan during the panning process. You don't want gold to escape out of your pan amidst some dry soil. Also, gold flakes that are bone dry will not always break the surface tension of water and sink to the bottom of your pan. This is more a problem when prospecting in the desert. We'll talk more about that in a later chapter.

Always deposit your gold into the vial with the gold pan underneath it. With the gold pan placed beneath your vial, should you drop a piece of gold, it will fall back into the pan where you can easily retrieve it.

After putting the gold into your vial, securely screw on the cap and put your vial in a safe spot. I learned this lesson the hard way. I had placed a large vial, about one-third full of nuggets, into my shirt pocket. When I leaned over to pan another batch of concentrates the vial rolled out of my pocket and disappeared under water where it was soon swept up by the raging torrent of the river. I had failed to button the shirt pocket, and the vial, heavy with gold, just rolled right out. So, if your pocket has a button, use it. If not, don't put your gold there! Here is a word of caution for wearers of tight-fitting, heavy denim jeans. If you put a gold vial in your front trouser pocket, then stoop down, you could crush the vial in your pocket. The result would be loose gold and shards of sharp glass in your pocket. I solve this problem by storing my gold vial in a zippered pocket of my prospector's day pack, which is usually lying near my panning spot. This way I also have access to my loupe—a small hand-held magnifier—which is also kept in the day pack.

Although gold vials come in a variety of sizes, you may recover nuggets too large to fit through the neck of even the largest vial. In this case, most prospectors tote along small plastic canisters — 35 millimeter film comes in them — to store and collect the larger nuggets. I always keep one of these film canisters in a day pack in which I carry various tools needed for prospecting.

Much of what you find while prospecting will be flakes and fine pieces of gold. These add up, over time, into profitable amounts of gold, but can be really tedious to separate from black sand and pick from the gold pan, so these fines, along with black sand, are washed into a coffee can for later processing. Fines are also useful for making nice pieces of display jewelry to wear or give as gifts.

Most prospectors accumulate black sand and fines from several prospecting trips for processing at a later time. There are a number of ways to process fine gold from black sand. One of my favorites is also one of the most enjoyable. I usually do this in summer while out on the patio. Here's what I do:

I fill my washtub with water and pull up a low-to-the-ground, folding beach chair. I have my radio playing some nice tunes, and a cold libation — all within easy reach. Near the washtub I place a vial for the gold, my collection of black sand, and my small brush for picking up gold. Then I spend a lazy afternoon soaking up rays of sunshine and panning out bits of gold from black sand. Occasionally, I will find a small nugget or two that somehow I missed seeing before saving the black sand in my coffee can.

I find the panning process quite pleasurable and relaxing, delighting in the process itself. I also enjoy sighting and picking out the bits of gold and placing them into my vial. I can spend hours at this and often do, simply because I like doing it.

Certainly, there are more time-efficient ways of separating black sand and gold than what I have just described — and we will get to them later, along with copper pans — but I just wanted to share my favorite method. Perhaps it will become yours too.

The first thing that a prospector does after reading the area for possible gold concentrations is to "sample pan" to determine just where the highest concentrations of gold are located. The successful prospector always sample pans an area *prior* to setting up any gold recovery equipment, such as sluices, highbankers, or drywashers. These types of equipment require that the prospector shovel loads of gold-bearing gravel into them for processing. If the gravel going into them is not gold-bearing, then the prospector has wasted a lot of time and done a great deal of shoveling for nothing.

Sample panning first, before setting up, saves the prospector a lot of time and work. If your sampling doesn't turn up any gold, simply move on to the next area where research has indicated gold may be found.

I see many prospectors step out of their vehicles, haul equipment down to a convenient location — usually not very far from where their vehicle is parked — and set up. Then, without even determining where, if anywhere in the area, the highest concentrations of gold may lie, they begin to shovel material. These individuals may work for a couple hours before the riffles in their equipment require emptying of concentrates. When they pan the concentrates, they get nothing! They may do more useless digging, or stand up, brush off, look around, and pack up, affirming, "Well, there's no gold here!" How would they know? They haven't performed the most vital step of successful gold prospecting — sample panning. Sadly, only a few feet from where they were digging unsuccessfully may lie a profitable concentration of gold flakes and nuggets — a concentration that will be located by the next prospector who does his or her sampling! A successful prospector always keeps in mind the gold prospector's Cardinal Rule, "Exploration should always precede extraction."

Sample panning requires a gold pan, a shovel, and a few small items that compose the "prospector's day pack." I carry a short "miner's shovel" and stow in my prospector's day pack a gold pan, a trowel, my crevicing equipment (more on this later), a geology pick, perhaps a topo of the area I'm in, a vial for gold, my wee brush, a pair of five-inch long tweezers, loupe for viewing fine gold, lunch, and a bottle of drinking water.

I take all this to a spot that my research has led me to suspect con-

The author samples bench gravels to check gold concentration.

tains gold. Then I find a nice spot to sit where I can survey the area and refer to my topo map. I then compare the area to what I know about reading the river and the properties of gold, then wander over to some of the spots that are likely to have concentrations of gold, keeping in mind how obstacles slow the flow of water. There I dig up some of the soil and pan it out, working my way systematically along, sample panning as I proceed. If no gold is found, I can move on to other areas, knowing I have tested thoroughly the possibilities.

Sample panning will pay dividends to the prospector who spends the time to find out where the gold is or isn't. Once concentrations of gold are located, it is time to return to your vehicle and collect your gold recovery equipment. On some occasions you may find concentrations of gold but decide that they are just not rich enough to bother with. Sampling allows you to locate several areas of concentration and then decide for yourself which you think is worthy of your efforts. Wearing my day pack and toting my short miner's shovel I can cover a lot of ground in a short amount of time, sampling as I go.

When I have covered a section of river, I hop into the car and drive

Sample panning yielded this prospector knowledge of where gold was concentrated in his area, and a nugget to boot!

to the next section. Usually, sampling doesn't take very long compared to several hours heaving barren gravel into your gold recovery equipment. Each area you visit will require more or less time spent sampling depending on the terrain you are covering and the number of obstacles that have slowed the flow of water. There are no hard and fast rules about how much time you should spend sample panning. The only rules are thoroughness and patience. You will be rewarded!

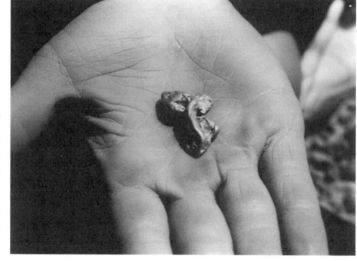

A dandy nugget like this makes a one-day prospecting trip a profitable one.

On a two-week prospecting trip to the Mother Lode of California, I spent four days sample panning a stretch of the North Fork of the Yuba River between Downieville and Sierra City. It paid off profitably because I spent the next week-and-a-half recovering major concentrations of gold along a rather doubtful section of riverbank. I suspect that this section of river had been bypassed by the majority of prospectors because of the difficulty of carrying down recovery equipment from the road. Due to the fact that I had arrived at this area not from the road above but by walking along the river, sampling as I went, I knew that approximately 150 yards upstream was a dandy place to tote equipment down from the road above, then carry it over to the spot where I'd located gold, and set it up.

When you find a spot where gold has concentrated, then and only then do you unload your recovery equipment, set it up as near your spot as possible, and get to work. If you sample pan, I guarantee that when you pan out your recovery equipment concentrates, you will find gold in your pan — every time!

Recovering Gold With A Sluice Box

To sluice for gold you will need:

1. Your sluice box.
2. A five gallon plastic bucket.
3. A screen of approximately $1/4$- to $1/2$-inch mesh.
4. A miner's or garden shovel.
5. A geology pick and/or a hoe pick.
6. Your day pack with pan, vial, loupe, tweezers, etc.

Most prospectors start with a sluice box once they've learned the art of panning. A sluice is relatively easy to carry and inexpensive. It doesn't take up much storage space and comes in several convenient sizes. There is even a little "back pack" size for the prospector who likes to hike into inaccessible areas, leaving the majority of prospectors behind. If you purchase a sluice without a handle, drill and bolt on a rubber-rimmed garage door handle for ease in carrying.

Most sluice boxes are made of formed aluminum to which is attached carpet, expanded metal screening, and metal riffles. Sluices can be purchased from a variety of prospecting stores, outdoor supply shops, and mail order catalogs. Expect to pay $25 to $100, depending on size. The size that I use is about three feet long by one

The author watches Ed Butler pour gravel into a sluice bed.

foot wide and costs about $50. The first third is composed of bare aluminum. The lower two-thirds contain removable carpet, screen and riffles. The "head" of many manufactured sluices is slightly flared. The corners of the sluice are bent outward and flattened so as to catch and funnel gold-bearing gravel into the sluice. The tail, or downstream side of the sluice, is where lighter, unwanted materials exit the sluice. Riffles are generally angled toward the downstream, or tail end of the sluice, so that gold falling out of the flow of water can secret itself in an area of little water movement. Although straight up-and-down riffles will trap gold, angled riffles maximize the "dead flow" area behind them, allowing more gold to accumulate with less water disturbance. Remember that most gold will settle out on the downstream sides of the riffles, just as occurs in nature with boulders and other natural obstacles.

Regardless of type, all sluice boxes operate on the same principle. Most contain a length of carpet to catch fine gold. The carpet is usually overlain with expanded metal screen which traps gold and heavier elements such as black sand. Over the carpet and screen, metal riffles are attached. The riffles catch and retain larger pieces of gold as they strategically alter the flow of water through the sluice.

The sluice comprises a mini-environment, the sole purpose of which is to separate unwanted lighter elements from the more desirable heavier elements. These heavier elements that remain in the sluice for later panning are referred to as "concentrates."

The sluice is set up at an angle in the water current, then anchored with rocks. The prospector then empties buckets of screened gravel into the head, or top, of the sluice. Water rushing through the sluice washes most of the dirt, rocks, and gravel out the tail end, while black sand and gold remain tucked behind the riffles. Rocks and gravel containing metallic elements will also remain with the concentrates.

The prospector should screen the gold-bearing gravel before sluicing. This is done so that large rocks falling into the sluice will not dislodge gold that may have settled behind the riffles. If gold is dislodged, it would most likely just be caught again behind the next lower riffle, but with numerous rocks falling into the sluice, water flow may be altered enough that gold could very well be lost back to the river. Also, large rocks banging down onto the metal riffles will in time mash and modify their shape, rendering them less efficient at catching and holding gold.

Another reason for screening the gravel is that few prospectors want to tote a bucket full of useless and weighty rocks to the sluice. Remember the prospectors axiom, "The more *dirt* you move, the more gold you get!" So remember, *rocks do not count!* Rocks are merely dead weight. Screen them out and leave them behind.

Once a prospector locates, through sampling, a spot that shows a fair amount of gold, the sluice is set up in a location with good water flow. The ideal spot for your sluice would be close to the bank so that you don't have to get your feet wet. Sometimes you just can't find such a place, so you can set up a few large stones to step on, or you can wear waterproof boots.

Place your sluice so that the head is slightly higher than the tail. You want at least two inches of water flowing rapidly over the bed of the sluice. Much less than two inches of water over the bed will be unable to wash out unwanted lighter material. More than two inches of water would be desirable.

More important than depth of water is speed of water flow. After a time, the prospector becomes adept at judging this. However, the beginning prospector should set up the sluice, and drop a predetermined number of BBs or metal shavings into the sluice head along with a handful of soil and gravel. What should happen, if your sluice is set at the right angle and with an adequate volume of water

A wing dam constructed with rocks angles more water through the sluice.

flowing through, is that your metal fragments should settle out behind the first few riffles while the soil and gravel should move rapidly out the tail end. A few rocks — especially if they contain iron — and black sand will settle out behind most of the riffles.

If your metal fragments pass on through and out of your sluice, the water flow is too fast. To remedy this:

1. Move the sluice to slightly shallower water.
2. Reduce the angle of the sluice.
3. Using rocks, reduce the water flow at the head of the sluice.

The flow of water through the sluice is too slow if your metal fragments and handful of gravel fall in and don't move anywhere. To remedy this:

1. Increase the angle of the sluice.
2. Move the sluice to slightly deeper water.
3. Move the sluice to a spot where the current is faster.
4. Build a wing dam of rocks to angle more water into the head of the sluice.

If none of these measure works, you will have to find another location along the river to set up your sluice. After the first few tries, you will become expert at eyeballing the water flow conditions and setting up your sluice accurately and effectively, without testing with metal fragments.

You can set up the sluice on shore and pour water over it. In this

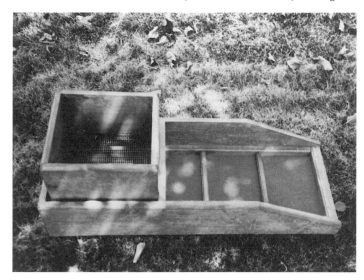

A rocker is used in areas of scant water flow.

case, set up the sluice at an angle so that it empties back into the river. Pour several cups of gravel into the sluice head. Then *slowly* pour water from another bucket into the sluice, washing the material through. This method is a bit tedious, but if you are onto good gold-bearing gravel in an area of little water flow, you can process the gravel using this method.

If you consistently work an area where water flow is poor, you should consider purchasing a "rocker." A rocker is similar to a sluice but is rocked sideways much like a baby cradle. Small amounts of water are then poured in to process the gravel. A combination of rocking and the addition of small amounts of water allows the rocker to process gold-bearing material. A rocker is a dandy way to process gold in areas where water flow is greatly reduced in warmer seasons.

You've set up your sluice near a spot that sample panning has indicated is rich in gold. Use your shovel to dig into your gold-bearing spot. A five-gallon plastic bucket should be conveniently close to you with your screen sitting on top. Place a shovel full of dirt on top of the screen. Shake the screen back and forth over the bucket, or you may rock the entire bucket if the screen is the manufactured type that fits securely into the mouth of the bucket. Soil, gravel, and gold will pass through the screen into the bucket. Rocks too large to pass through the screen will remain on top. Be sure that all the gravel falls into the bucket and not around the outside of it or you could lose some gold.

Before you discard the rocks remaining on top of your screen, do a visual check to be sure that no gold larger than $1/_4$-inch is sitting on top of the screen. Just to be extra sure, use your hand to move the rocks around.

A sad tale, both to tell and to hear, occurred to the father of a prospector friend of mine. Her dad was prospecting around Downieville in Northern California and he forgot to check the screen top before discarding rocks. His last sight of a beautiful large nugget was its flight through the air amid the rocks he had just tossed off. Although he spent the rest of the day looking for the nugget near where most of the rocks had landed, he was never able to locate it.

Remember to always discard your rocks at a spot where you do not plan to dig, or have dug and found nothing. There is nothing more irritating, or time-consuming, then to have to remove a load of rocks from the next spot where you want to dig. Either toss the rocks off your screen into the river (provided you are close enough), or drop them where you have already dug out, and exhausted, any gold-bearing gravel.

If you did your sample panning thoroughly you will know a spot nearby that has no appreciable gold where you can dump your rocks. Dump them only as far from you as you can easily toss them from your screen. The idea is to spend time digging and dumping gold-bearing gravel into the sluice, not toting about screens full of rocks. So quickly dump those rocks (after scanning for large nuggets, of course) and get back to work!

Once your bucket is full of gold-bearing gravel — or as full as you can comfortably lift — carry your bucket to the sluice. Begin to shake the contents of your bucket into the top (upstream end) of your sluice. Do not dump the bucket all at once. You want to shake the contents into your sluice using a "measured shaking" motion so that all the material falling into your sluice is completely wetted. Dry material will ride through the sluice on top of the water and gold will not settle out into the riffles. Remember that gold must break the surface tension of the water in order to sink. Therefore, gold, especially the fines, must be wet. Also, if you dump the entire bucket at once, it will be more dirt than the sluice can process.

After several buckets have been processed by the sluice you will notice that the spaces between the riffles have begun to fill with ma-

Cleaning sluice concentrates into a bucket for later panning.

terial. When these spaces are full, nearly to the top of the riffles, it is time to take up the sluice and clean it. To do this, just move the rocks that anchor it, being careful to keep the concentrates from washing out of the tail end of the sluice. Raise the sluice, keeping it as level as you can, so as not to spill your concentrates, then upend the whole thing into your empty plastic bucket. Sluices that are 12 inches or less in width will slide perfectly into five-gallon buckets.

Using your gold pan, or another bucket, pour water over the sluice bed, washing the concentrates into the bottom of the bucket. Continue to add water, washing the concentrate free, until your bucket is full of water. Loosen the carpet, expanded metal screening, and riffles while the sluice is in the bucket. Pull the sluice bed free of its screen, riffles, and carpet, wash away any concentrates adhering to the sluice bed, and set it aside. Rinse the riffles and screening and set these aside. Carefully scrub the carpet free of any clinging material. You may use a small scrub brush for this or merely scrub the carpet against itself, then remove it and set it aside.

Now carefully and slowly pour the water in the bucket into your gold pan. There will be more water in the bucket than the pan can hold; that's okay. Just let the water overflow the edges of the gold pan — *until the concentrates begin to pour into the pan from the bucket.* Pour just enough concentrates into the pan for comfortable panning, making sure that they do not pour over the side of your gold pan.

Pan the concentrates as previously described and place any gold that you find into your vial. If any concentrates remain in the bucket, pan these next. You will probably find that some of the concentrates adhere to the bottom and sides of the bucket. Toss a few handfuls of water into the bucket with one hand while the other hand holds the bucket in position over your gold pan. Tossing water up into the bucket will cause the remaining concentrates to flow out into the gold pan. Just be careful that all the concentrates fall into the pan and not outside of it. The last of the concentrates in your bucket are likely to contain the most gold. Remember that gold has a tendency to sink to the bottom of whatever it is in, and most of it will be found at the bottom of your bucket with the last of the concentrates.

When you find that the riffles of your sluice seem to be filling too rapidly with material, and that you have to break it down and process concentrates after every couple of buckets, it may be because the flow of water passing over the riffles is too slow. Too many lighter materials are remaining in the sluice. Adjust the angle of the sluice or increase the flow of water. Breaking down and cleaning a sluice too often can be tedious.

You will want to save the black sand from each panning into a coffee can or other plastic, lidded container. It is unlikely that you will want to waste time picking out minute flakes of gold, when you could be processing more gold-bearing gravel through your sluice. So save your black sand and fine gold for processing at your leisure. Just pick out the big flakes and the nuggets, adding these to your vial, and save the small stuff for later.

If your concentrates contain gold, reassemble your sluice and re-anchor it with rocks. You will want to keep digging, screening, and carrying gold-bearing gravel from your spot to your sluice as long as you are getting gold.

When your concentrates come up empty of gold, or your spot seems to be lessening in gold concentration, you will want to begin digging in another spot where sampling has indicated gold is to be found. This may necessitate setting up the sluice in a place that is more convenient to your new location.

Remember not to overwork yourself. Take rest breaks when you need them and be sure to stop and enjoy a lunch period. It is tempting, if you are into great gold-bearing gravel, to bypass breaks and lunch. I have done so and suffered the consequences. In the long run you will be glad that you stopped for lunch, at least. A midday lunch break will go a long way in providing your body with the fuel it needs to keep going for the rest of the afternoon.

I have been known to dig, sluice, and pan by lantern light when into really rich gold-bearing gravel. And to get up before dawn to continue processing this gravel, if time at a location is necessarily short due to other commitments. The lure of gold is such that if you find yourself in an especially rich area, you will be amazed at the energy and endurance that your body will manifest.

The mere sight of gold is very energizing to most prospectors. You will be having so much fun and satisfaction recovering profitable amounts of gold that time will pass and you will not notice any tiredness or desire to stop. Just remember not to overdo it. No amount of gold is worth risking your health. Take breaks when you need them. Eat to keep your body fueled. Be sure to drink ample fluids. Be good to yourself, and your prospecting will be healthful and enjoyable.

In an area open to public access, leave your shovels and buckets in or near the hole that you are working as a sign to other prospectors that you have a temporary "claim" on that spot.

Most prospectors will honor your temporary claim and leave it, and your equipment, undisturbed. Be aware that there are a few unsavory souls who will not only work your spot, but may abscond with your equipment if you are not around. Hence, when onto good gold, you too will likely get up before dawn in order to be back at your spot and working it before others awaken. You may also wish to move your camp closer to where you can keep an eye on your spot and your equipment. Most prospectors set up camp in the vicinity of the spot that they're working.

When you are onto good gold, or find a dandy nugget in your pan, you will not want to advertise the fact to any other people — besides your companions — who may be in the area. Act nonchalant and, if asked how you are doing, shrug and tell a little white lie. If you let other prospectors know how well you are doing, they will soon be working right on top of you. You may even want to dump your black sand, along with all the gold into your coffee can for panning out later, when no one is around to watch you.

Many prospectors enjoy panning out in camp in the evening — by lantern light — while sitting around the campfire with their companions (and a warming libation). Sluicing is a great and inexpensive way to recover profitable amounts of gold as long as the water flow is sufficient and your gold-bearing gravel is not located too far from the sluice.

Researching Gold Sites

You now know the basics of gold prospecting. You know where along a watercourse to find gold and how to select and prepare a gold pan for use. You know how to sample pan an area for gold concentrations, how to sluice and process your concentrates, as well as how to remove gold from the pan into your vial.

Now I want to talk to you about another very important aspect of gold prospecting. We skipped this point in order to get to the basic prospecting procedure because most people are eager to get to the actual techniques — or bottom line — of gold prospecting. But now that we have covered the basics, it's time to make a very important point, and that is RESEARCH! Research, research, research! This cannot be stressed enough. The most successful prospectors are those who do their research.

A well-known fact is that gold is found by prospectors today where it had been found previously by prospectors yesterday. Although it is remotely possible (and I mean remotely!) to go out prospecting and find a heretofore undiscovered gold deposit in an area where gold has never been found before it is highly unlikely that you will be the one to do it. Gold deposits are located in our modern age by mining professionals, using aerial and satellite survey methods and extensive ground sampling. There are certain types of geologic processes that result in gold deposition and in this day and age, nearly all significant prospects have been checked out by professional geologists or commercial mining companies. That leaves little to be newly discovered by the rest of us.

My prospecting partners and I are always finding *overlooked* gold deposits — and so will you — in historic gold-bearing areas, but never have we gone into an area and found gold in a spot that was not known to have produced some gold at some point in its past.

Over the last century, geological surveys have identified almost all of the gold-bearing areas in the world. For this reason it is unlikely that the recreational prospector will ever hit upon an area that has never been identified as potentially gold-bearing.

On the positive side, gold is constantly being concentrated and transported within known gold-bearing areas. Also, due to the processes of erosion and weathering, gold deposits are continually being exposed at the surface. What may have been buried, or inaccessible, a few years ago, may now lie exposed due to a variety of weathering and geologic processes. These are the deposits that recreational prospectors are seeking and finding. Finding these types of deposits and concentrations is the edge that every recreational prospector is hoping for, and what recreational gold prospecting, for fun and profit, is all about!

As stated previously, the contours of a river change over time. Runoff from rain and snow varies constantly from season to season and moves gold from one spot to another along the length of a watercourse. So, from one season to the next, gold may not be found in the same place along a river. The concentrates will relocate and reconcentrate time and again within an area assisted by periods of heavy seasonal runoff. This is where reading the river combines with your research as an invaluable aid to successful prospecting.

Before reading the river, always do research. Research is done in libraries — through books and magazines on the subject of recreational prospecting, by looking up old newspapers and reading them for information on historical gold finds, and by talking to other prospectors, or by scanning topographic maps for old placer and lode workings.

Some areas are famous for historic gold finds; even non-prospectors have heard of them. Who hasn't heard of Northern California's Mother Lode and the 1849 Gold Rush, kicked off when John Marshall espied gold in the millrace at Sutter's Mill?

Not surprisingly, much of the gold recovered by California prospectors is still found in this historic gold-producing area. Less well-known are the other gold-bearing areas in both northern and southern California. Too, almost half of the states in the U.S. have known gold deposits. Don't forget that parts of Great Britain, Australia, Canada, South America, Africa, and the Philippines contain gold deposits. In all of these places are historical gold-bearing sites where recreational prospectors today are recovering profitable amounts of gold!

Research allows the prospector to discover these lesser-known areas and tap the gold resources that have been accumulating and revealing themselves anew since their initial discovery decades ago.

There are numerous and interesting books, magazine articles, and maps describing the history, location, and production of gold-bearing areas. Much of the material originated years ago, but newer techniques and equipment, as well as prospecting information, are readily available.

The California Division of Mines and Geology, or analogous agencies in the various states, publishes bulletins detailing the mineral resources of the state. These can be found in used book stores, in libraries, or in the research departments of colleges that teach geology. It is possible to mail-order bulletins from government agencies, as well. Gold is one of the mineral resources listed in bulletins of this type. For most of the listings in bulletins, range and township numbers for topographic maps are listed, which are a great help in locating potential prospecting sites.

The U.S. Geological Survey, a division of the United States Department of the Interior, can be of help in researching historic gold areas. The U.S. Geological Survey publishes topographic maps which show many old mining sites. The address of the U.S. Geological Survey, as well as ordering information, is listed in the last chapter of this book.

Let's Go Gold Prospecting...On A Budget!

Many of us who have a yen to go gold prospecting just don't have a lot of money to spend to get started. Gold prospecting equipment can be quite costly, particularly motorized or electronic equipment. But the good news is that gold prospecting does not have to cost an arm and a leg. The even better news is that expensive equipment will not make you a successful prospector if you lack the *basic* skills. What makes a successful prospector is good research, skill in reading the river, and careful, thorough sampling techniques.

Initially a prospector should plan to lay out between $5 and $7 for an inexpensive, drop-bottom steel pan. Plan to spend a few bucks more if you want plastic. Throw in a long-handled garden shovel from the garage. For free, you can pick up a couple of five-gallon plastic paint buckets from behind paint supply stores or professional painting outlets. The same type of plastic bucket is also used to ship and store syrup, cleaning solutions, or other liquids, and may be found behind large institutions such as hospitals, some industries, and restaurant supply companies. Just be sure to clean them thoroughly before use.

In the case of used paint buckets, you don't want to introduce flaking paint into the natural environment. If your buckets were used for food storage you could be risking an invasion by ants or other pests if they are not cleaned out. These same buckets that you can find for free with just a little searching will cost around $5 to $10 if purchased from geological or hardware supply companies.

An item you will eventually want to acquire is a hoe pick for breaking up large sections of earth. Many of you may already have this item in your garage. If not, it is possible to purchase a used one at garage sales. New, a hoe pick costs between $35 to $40.

A must in all types of prospecting is the rock pick, also called a geology pick. There is one on the market called an economy pick that costs under $10, but this pick will dull quickly. It is better to pay for a good quality steel rock pick that will give you decades of use. In my nearly 20 years of gold prospecting, I still own and use my first purchased pick. These can be found at most hardware stores, prospecting supply shops, rock shops, and through many lapidary or geology supply catalogs. You may also purchase them used by watching the classified sections in newspapers for garage sales and estate sales, or through mineral club members who are selling off their equipment for one reason or another.

The rock pick can be used in place of the hoe pick, but will loosen a smaller volume of earth with each swing. The rock pick is useful for working in tight quarters, loosening packed soil for sampling, and for cracking open ore samples.

Many first-time prospectors purchase a sluice box new, but it is possible to find used ones at flea markets and swap meets. Large gem and mineral shows, like the outdoor show held in Quartzsite, Arizona, each February, is a good place to pick up used, or new, prospecting equipment and tools.

You can easily make a sluice box. If you are a metal former you are on the way to creating a low-cost, lightweight and inexpensive sluice. However, if you have even minor woodworking skills, or none at all, you can make a sluice out of wood for a minimal cost. I made my first sluice from wood. I must admit that after sitting in water all day, it was heavier than any aluminum sluice you could purchase or make, but it did the job for a couple of years. Then one day while prospecting I found an aluminum sluice box bed — minus the carpet, expanded metal screen, and riffles—that someone had discarded. Using auto/marine silicone cement, (waterproof, of course) I cut and glued in one-by-one-inch oak riffles. Between these I glued in four-inch-wide strips of ridged rubber matting, and had a sluice that took me through the next five years of prospecting.

Finally, one day I purchased a three-foot-long aluminum sluice box for about $28 and gave both of my old ones to friends eager to try gold prospecting. The same three-foot sluice that I purchased 12 years ago costs between $50 and $60 today.

For those who want to make a sluice, I have included a helpful diagram. In this homemade version, rubber matting takes the place of carpet for catching fine gold. With a homemade sluice there is nothing to disassemble. You merely rinse it thoroughly in your bucket by pouring water over it with a gold pan or bucket. It is much faster to clean than commercially manufactured sluices, but be prepared for it to be heavier at day's end, especially after it has sat in a river for hours absorbing water.

Before we leave the subject of sluices, I want to pass along a hint for success. Should you purchase a sluice, you will notice that the first third of the sluice is bare metal. This bare metal wastes a third of your potential gold recovery surface. To remedy this, a ridged piece of rubber matting can be cut to fit and glued onto the bare metal. I call this addition to a purchased sluice a "nugget trap." This is because nuggets and large pieces of gold will tend to fall out into the ridges of the matting due to gold's high specific gravity. Also, many small flakes will fall out onto the matting.

By installing a nugget trap you will be able to give a quick visual check, even through running water, to see if you are still running gold-

A "nugget trap" of rubber matting maximizes gold recovery and minimizes cleaning of the sluice.

bearing gravel. Without a nugget trap you cannot tell if you are currently running gold-bearing gravel unless you take the sluice up, dismantle it, and pan out its concentrates. The only thing besides gold that will fall into your nugget trap is black sand and garnets. Most of the gravel will fall out farther down into the areas of the metal riffles. This is due to the lighter specific gravity of most rocks and gravel. The nugget trap is an easy-to-install improvement that will save you time and effort. It also increases your gold recovery surface by about 33 percent. The best rubber matting for making a nugget trap or lining the sluice bed between the riffles of homemade sluices comes from under the carpet in the back of Volkswagen vans.

The best place to acquire this rubber matting is from auto junk yards. Check out the matting in the backs of VW buses by pulling it out and turning it over. The bottom side should be covered with one-inch squares. The sides of the squares are about $1/_{8}$th of an inch high. If you are unable to locate VW matting, hardware stores sell rubber matting by the roll. You just cut off the amount you need. Be sure the matting has ridges that will act as riffles to trap gold. Also, be sure to position the ridges on your matting so that they run parallel to your riffles.

Keep a pair of tweezers clasped to the side of your sluice. Use these to remove any nuggets or large flakes that settle into your nugget trap. You will not be able to remove them with your fingers; as soon as you try to, the gold will swirl away and down to a lower sec-

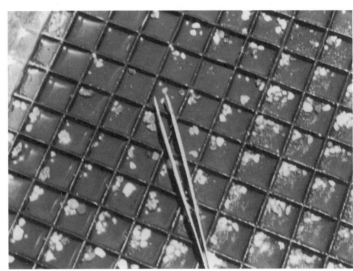

Tweezers are essential for picking gold out of the sluice's "nugget trap."

tion of the sluice. Your fingers, dipped into the water flowing over your sluice, are large enough to alter the current. Water eddying around your fingers will cause gold to be displaced. Tweezers are too thin to cause this problem and you can easily pick out gold without having to take up the sluice.

Around your home are a few more items that will make your prospecting an inexpensive affair. These simple tools, which are found in nearly every home, are used when "crevicing" and "sniping" for gold. For these two techniques all you need are a gold pan, a vial for gold, and a few simple tools that are probably residing in your garage and kitchen.

Crevicing For Gold

Crevicing for gold is one of the exceptions to the rule that a gold pan is only a sampling and cleanup tool. Here, a gold pan **is** your main gold recovery tool. Crevicing is a simple technique that is best used in the spring, just as the water level in rivers and creeks begins to drop. Soon after the snow has melted and the water level begins to fall — often as early as late March or early April — I can often be found crevicing in the exposed bedrock of rivers for the gold they contain.

Runoff from rain and snowmelt increases the volume of water in a river or creek bed, causing the water to rise. Later, when the water level begins to drop and the volume lessens, bedrock is exposed. Bedrock is filled with cracks, crevices, and cavities which trap and hold gold nuggets and flakes. These crevices in bedrock are also packed with sand, rock, and gravel. Using a variety of simple tools one spoons, scrapes, brushes, and suctions these crevices out, emptying the contents into a gold pan for subsequent panning.

Many prospectors crevice for gold when the rivers are still rushing too swiftly to allow for sluicing or dredging or before dredging season officially starts. Crevicing is a great activity for people who don't relish the labor of sluicing. You simply sit upon bedrock (you may if you wish provide yourself with a pillow), emptying and panning the contents of the crevices. You remain dry and relatively inert until it's time to get up and settle down by another crevice.

The tools you need for crevicing are a gold pan, a vial for gold, a small spoon, a long-handled screwdriver, a whisk broom, a paper plate, a long bladed garden trowel, and a turkey baster. You may want to add some thin stiff wire, such as piano wire, to get into really narrow crevices. You'll probably want to wear gloves to protect your hands from inevitable scrapes with bedrock.

To crevice for gold, pick a crevice that is packed tightly with sand and gravel. An empty crevice is likely to mean that another prospector has beaten you to it. Even so, many prospectors fail to completely clean out the crevice, which means that they have also failed to get all the gold. Gold sinks to the bottom of anything, and if the crevice is not completely cleaned out, you will miss most of the gold. Many prospectors stop crevicing when they reach a fine white clay. This is a mistake; embedded in, and under, this clay is more gold. Using your

Day pack with gouging and digging tools for crevicing and sniping. Also shown: gold pans, loupe, tweezers.

screwdriver, bring up bits of the clay and place it in your gold pan. Rub and wash this clay until it is completely dissolved. Then you can pan it out and recover the gold that others left behind.

Your spoon, screwdriver, and wire are used to lift and scrape out sand and gravel. A whisk broom or stiff-bristled paint brush can be used to sweep the bottom of wide crevices. Sometimes crevices are too deep to permit access to the bottom. In this case, use a turkey baster, or bulb snifter (for adding water to a car battery) to shoot water into a crevice. This dislodges gold and causes it to rise briefly. Immediately release the bulb, allowing water and gold to be sucked up. Plastic squeeze bottles with long, cone-shaped siphons can be used also. Squeeze the bulb or bottle contents into your gold pan to be panned out with the rest of the contents of the crevice.

Crevicing is one of the simplest ways to find gold, and is also the least work. The prospector remains dry and fairly stationary. If I am on a long prospecting trip, I may sluice one day, crevice the next in order to rest sore muscles, nugget shoot the day after that, then start all over again with sluicing. However, if I am into a glory hole that can only be recovered by sluicing, I will continue sluicing until the deposit is exhausted. I like to employ several methods to give diversity and interest to the days I spend prospecting. By using a variety of methods in one area, I maximize my gold recovery. There are advantages to each recovery method, which you will discover as we go along.

Sniping For Gold

Sniping — searching of underwater crevices for gold — is the other exception to the rule of the gold pan. Here again a pan is your main gold recovery tool. Sniping can be done while remaining fairly dry on bedrock too, or by wearing a wet suit and using a snorkel.

For dry sniping you use an easy-to-make tool called a "prospector's glass." Using the prospector's glass you peer into the water looking for crevices just beneath the surface. Using crevicing techniques, you remove material from underwater crevices into your gold pan. You will have to adjust your techniques somewhat to compensate for the speed of the current, so that material flows from the crevice into the gold pan that you hold beneath the water. Your bulb snifter, or turkey baster, comes in handy for sucking material from underwater crevices, too. Occasionally you will espy a nugget wedged into a crevice. Use a pair of tweezers to pull it free and have your gold pan positioned to catch it should you fumble and drop the nugget.

The summer months, when the temperature of the water has warmed and the days are hot, are a good time to immerse yourself in the river for sniping, providing the current is not too swift. Wearing a diving mask and snorkel you can get at more underwater crevices. The advantage of sniping over crevicing is that you will have both hands free to work the crevice.

To make a prospector's glass you need:

1. A tall juice can.
2. A one-millimeter-thick sheet of styrene plastic.
3. A coping or jeweler's saw.
4. A garage door handle.
5. Screws and nuts to attach the handle.
6. Waterproof silicone cement.
7. Lead-free, rustproof spray paint.

Remove both ends of your juice can and remove the label. Wash and dry the can thoroughly. Spray paint the can inside and out. This will eliminate glare from shiny, unpainted metal. Spray paint your handle to match. Place the can on the sheet of styrene plastic and trace a circle around the inside of the can onto the plastic. Styrene can be purchased from hobby or craft shops, or check your phone directory under "plastics."

Use your saw to cut out your circular shape. Use a grinding wheel or sandpaper to smooth the rough edges of the plastic. Squeeze a ring of silicone cement on what will be the bottom side of your styrene lens. Insert the lens into the bottom of the can. Squeeze another ring of cement to the inside or top edge of the lens where it meets the can. Smooth the cement on both sides of the lens with your finger to fill any gaps and pockets. Silicon cement should not come in contact with skin, so wear a rubber glove or insert your finger in a plastic baggy to prevent contact with the cement. To remove any cement you may have dropped onto the lens or gotten on your skin or work surface, you can use rubbing alcohol applied with a cotton ball.

Allow the cement to cure for 24 hours. Then immerse your prospector's glass into water to test for leaks. Patch any leaks and cure for another 24 hours. Drill holes in the side of the can and attach the garage door handle. Squeeze a bead of cement on the inside of the can around the screws and nuts to keep water from seeping into the can around the handle. You are now ready to go sniping for gold!

Dowsing For Gold

Dowsing is an ancient technique that has been used to locate water, oil, gold, lost objects and people. There are many theories about why and how it works. I believe that no occult forces are at work in dowsing, only unsolved mysteries of the mind. A study done some years ago on the mechanisms of dowsing appears to confirm my opinion. The study was done by a group of scientists and other professionals belonging to the International Research Council of the Rosicrucian Order, AMORC, a worldwide organization devoted to the study of art, science and mysticism. Their results indicate that minute, unconscious muscular contractions (motor automatisms) in the wrist of the operator cause the pendulum or rods to move. These contractions were so subtle that the operator was unaware of them and had no control over them.

What is not known for certain about dowsing — and there are various theories — is the connection between that which is hidden in the ground and the mental stimulus that causes the muscular contractions in the vicinity of the object being searched for.

It is not my intention to try to convince anyone that dowsing is a sure technique for finding gold deposits; it isn't. I have had limited success using dowsing to locate gold deposits, although underground water and buried electrical cables are relatively easy to locate. Gold is more difficult for some reason. My only aim here is to introduce you to a technique that you may like to try for yourself.

Dowsing is not one of the techniques I employ most of the time. I have used dowsing when all other avenues of finding gold deposits in an area have proven futile. Here's how it works. There are two types of dowsing. The first technique is called "map dowsing" and uses a pendulum to indicate areas on a topographic map that are likely to contain gold. A pendulum can be made out of a string or length of chain about six inches long. To the end of the string is attached a locket, crystal, stone, bullet, or some other heavy object. Rest your elbow firmly on a table, your thumb and index finger firmly, but gently gripping the string end. Allow the pendulum to hang motionless, suspended over the map. Hold in your mind an image of what you wish to locate — in this case, gold. Mentally ask the question, "Are gold nuggets to be found here?" If the pendulum begins to circulate in a counterclockwise motion, the answer is "no." If the pendulum begins to circulate in a clockwise motion, that indicates a "yes" answer. When the pendulum remains still, it usually means that you should either rephrase or clarify your question. It can also mean that you are not thinking clearly about the object of your search or your question.

It is advisable to practice using the pendulum by asking simple questions to which you know the answers before you attempt map dowsing. You must become familiar and at ease with using the pendulum. Also, you must not consciously move the pendulum yourself; you must allow it to move on its own, apart from your conscious control. Remain relaxed.

When map dowsing, mark the spot or spots that the pendulum has indicated where gold may be found. Visit the site and determine if this is so. You may find that you have a talent for this mysterious and little-understood method of locating gold.

The other type of dowsing requires dowsing rods. These can be made by straightening two wire hangers and reducing their length to 20 or 22 inches. Bend them in half so that you have an "L" shape. Cut a plastic drinking straw in half. Placing one end of each "L" into a straw half. This allows the rods to swing freely.

To dowse, lightly grip the straws, holding the rods level and parallel to each other in front of you. Visualize strongly the object you desire

Map dowsing for gold with a copper pendulum.

to find. Do not allow other thoughts to intrude. As you get closer to the object the rods will begin to swing towards each other. If you are getting farther from the object they will swing apart. When you are standing over the object you are seeking, the rods will cross over one another. Dowsing rods are used to locate objects when you are at a site; map dowsing at home can locate potential prospecting sites.

The keys to success with dowsing seem to be an open mind and good visualization or imaging skills. Dowsing seems to be a mental ability. If you are convinced it will not work, it won't. If you can visualize well, you will likely succeed with dowsing.

If you have trouble visualizing, you may want to try a "witness." The witness is a sample of the object you are looking for. Since you are searching for gold, you would carry in your hand, or upon your person, a gold nugget while you conduct your dowsing. A witness may be used with either rod or pendulum dowsing.

In the next chapter we will examine other types of recovery equipment such as sluices, highbankers and drywashers.

Other Equipment & Methods

After mastering the use of gold pan and sluice, the basic tools of recreational gold prospecting, it is time to consider the other types of gold recovery equipment on the market. While out in the field you may have observed other prospectors using dredges, highbankers, or drywashers. You were curious, but possibly a little hesitant when you learn that this equipment can cost hundreds of dollars.

Do-it-yourselfers can save money by manufacturing much of their own equipment. Kits can be purchased and assembled. Plans and diagrams can be mail-ordered so that you can build much of what you use in the field.

When you begin gold prospecting, you can be so galvanized by the sight of gold that you go a little wild. Don't run out and buy expensive prospecting equipment that may eventually languish in your closet or garage unused. Too many people do this, then their interest moves to other areas, and the expensive dredge is shoved up into the garage rafters to gather dust along with the ski equipment, the golf clubs, and the exercise bike. Take your time. Find out if recreational gold prospecting is going to become a long-lasting part of your life.

Before you make a major investment, allow me to pass on a little wisdom. Start out with pan and sluice. Read the river. Do your sample panning. Accumulate your gold finds. Study and read all you can about the various types of recovery equipment and prospecting methods. Go to gold prospectors' conventions. Dealers of most major brands of recovery equipment will be on hand to give demonstrations and answer questions. But don't be swept away by excitement, the end result of which may be the spending of money set aside to pay the mortgage.

You may be tempted to rationalize your expenditure by thinking you will recoup it all with gold found with this new, exciting — and expensive — piece of equipment. Rest assured that you will be able to recoup your expenditure, if you have first mastered the basics, and if you maintain a long-term interest in recreational gold prospecting. Time and the consistent development of your skills will get you past the "ifs."

Also, in your inspection of recovery equipment, glean a full and complete understanding of the principles of its operation. Question the manufacturers. They will be happy to explain how their equipment works. After a thorough inspection and understanding, you may find that you have the skills necessary to build a similar piece of equipment. Too, a thorough understanding of how a piece of equipment works will allow you to compare the different products and pick one that will work best for you.

When you can accumulate profitable amounts of gold with your sluice, you will have the skills necessary to accumulate profitable amounts with other pieces of equipment. If you are not recovering much gold with your sluice, you need to work on developing your patience and thoroughness in research, reading rivers, and sample panning.

Your gold prospecting should pay for itself. When you have recovered enough gold with your sluice to pay for a dredge, then, and only then, is the time to buy one. Not before. By allowing your prospecting to pay for itself, you will have ample time to study various techniques of prospecting and the related equipment, and decide if that method, or methods, appeal to you.

Then, when your dredge, paid for with gold found by sluicing, has accumulated enough gold to pay for a drywasher — and after you've investigated desert prospecting sufficiently — go out and purchase the drywasher. Your conscience will be clear, you will know that you have the wealth (in gold, or cash by selling the gold) to afford your new investment (and it will be an investment!), and the "if" of "what if I lose my interest," will no longer be a matter of concern.

Dredging for Gold

Dredging is one of the most popular forms of recreational prospecting. Those who enjoy diving will enjoy dredging. Of course, you don't have to dive to dredge. Many prospectors dredge spots that are no deeper than their knees. They wear rubber boots or fishing trousers to keep their water-immersed parts dry. For deeper spots one can use a mask and snorkel. The really deep spots require full scuba gear. Those who like to prospect by themselves usually purchase small, "backpack" type dredges. These are easily toted, set up, and operated by one person. The cost of a backpack dredge can run $600.

An eight-inch dredge can cost $12,000, or more, when new. Dredges are an investment, to be sure, but they can be an investment that keeps on giving when one has learned well the basic skills of prospecting. Check the classified sections of prospecting-related magazines for used equipment. You can get some really good deals on dredges by looking for used equipment.

Dredging in deep water can be dangerous unless a little caution is used. The danger of dredging comes if you are not careful when suctioning gravel from under large boulders. Without care, this can become a problem if you fail to notice that you may be undercutting a large boulder, causing it to shift, possibly trapping you underwater.

Dredging deep water requires a buddy system. One partner stays on the dredge while the other dives. That way if you get into trouble there is someone to help, or go for help. A few prospectors have drowned when they got into trouble underwater. Even with a buddy there can be trouble should he be unable to move a boulder pinning the diver, or if the diver's air supply runs out before the partner can summon authorities with the equipment to help.

Work out an emergency strategy. Never stay under until your air runs out. Have extra oxygen tanks available in case of trouble underwater. As with most forms of prospecting, a little caution, along with a constant awareness of what you are doing, will prevent most accidents. There are profitable amounts of gold to be found around and under large underwater boulders, hence the reason that a few prospectors have gotten themselves into trouble. Just be careful and you will have profitable, fun prospecting trips.

Take turns going under water to prevent over-tiring. Exposure to cold water for long periods of time can be debilitating. When material stops coming up the hose, and over the sluice bed, there is probably a reason. Check it out. Of course it could be, and usually is, only a rock jam. On dredges that pump their own air supply, when the motor stops, or runs out of fuel, the air supply stops. Be aware of this. Your diving buddy's well-being depends on it.

Develop a means of communication for letting each other know that all is well, or not well. Talk over the things that could go wrong and develop a set of signals that all partners are familiar with. Communicate! This is the most important key to success and safety. Now that we have stated the most obvious dangers, and talked about how to avoid them let's talk about this most popular — and profitable — method of gold prospecting.

For dredging you will need:

1. **Wet or dry diving suit.**
2. **Diving mask.**
3. **Weighted belt.**
4. **Crevicing tools.**
5. **Air tanks or snorkel.**
6. **Rock pick.**
7. **Swim fins.**

A dredge is basically a sluice box that is floated upon pontoons or inflated inner tubes. A pump is attached to draw water from the river which is then directed over the sluice bed. A long hose is used by a diver to suction — to literally vacuum up — the material composing the bed of the river, or in some cases, a lake or natural pool bottom. The material is sucked up the hose where it is deposited, then washed over the sluice bed. The principle is the same as that of the sluice, only with the addition of more power! This power is usually provided by a gasoline engine, which operates the water pump and suction mechanics of the dredge.

The size of a dredge is usually determined by the diameter of the hose that is attached to it. The larger the hose diameter, the bigger the dredge and motor, the more gold-bearing material that the dredge can process — and the higher the purchase price!

Larger hose diameter means that bigger rocks and more gravel will be drawn up the hose and over the sluice bed. The more gravel that is processed, the greater the potential for gold recovery. The ability to suction up larger rocks means that the diver won't have to move them out of the way of the hose. Of course, the idea is not to suck up rocks, but gold-bearing gravel. Rocks are just part of what the suction dredge will pull up.

Most dredge prospectors pack along a metal rod just in case a rock becomes lodged in the hose. While the aperture of the hose nozzle is usually a bit smaller than the hose itself — to help prevent a stone from entering that would cause a jam — some rocks will go in easily but jam further up when they turn sideways. In this case the dredge needs to be shut down while the offending stone is removed. Jammed hoses are one of the minor, but bearable, problems encountered while dredging. The responsibility for preventing a rock jam falls to the diver, who tries to spot these possible troublemakers before they go up the hose. The dredge operator monitors the motor operation and the flow of material across the sluice bed. Should any large nuggets be spotted, the operator can shut things down and retrieve them, being sure to let the diver know that they're into a good run of gold. The diver may not have been able to see the nuggets among the volume of sand, gravel, and rock.

One- and three-inch dredges are fairly small capacity dredges. Those with six and eight inch capacities begin to assume the dimensions of small barges upon which the operator can stand or sit.

The backpack model dredge is carried attached to a pack frame worn on the operator's back. These small dredges can be carried into remote areas and used in shallow water spots. Most users of backpack dredges merely don knee-high rubber boots and may or may not completely immerse themselves. In the summer months when the water is less chilly, many will use a snorkel, enjoying a cooling dip while vacuuming the bottom for auriferous (gold-bearing) gravel or suctioning out underwater crevices in bedrock.

Most other dredges are carried to and from the prospecting location in the bed of a truck, sometimes requiring two or more people to carry them to the river. If the way down to the river is really steep, most prospectors will find a way to lower the dredge using ropes, then climb down after it.

Dredges come in "back pack" sizes as well as behemoths like this 8-inch model.

Dredges are anchored mid-river by lines running to both sides of the river in order to secure them against the current. When the river gravel in the vicinity of the dredge has been worked, the dredge is moved and re-secured in the new location. Many dredge operators work their way upstream, allowing the detritus to fall behind into areas already worked.

Dredge operators are looking for profitable concentrations of gold called "paystreaks." Paystreaks occur when strong mid-river currents move gold. This gold encounters obstacles, such as boulders, or protrusions of bedrock, and falls out of the flow. To recover paystreaks, dredge operators suction up the unconsolidated material which contains gold. If the layers of unconsolidated material are not too many feet thick, the desired outcome is to reach bedrock. At bedrock, the diver can suction the crevices and cracks in bedrock for rich concentrations of gold secreted there. It is common knowledge that the most gold will be found at bedrock.

Some of the larger dredges come equipped with their own trailers. Many come with air pumps for supplying the divers with air via an air line. Some really sophisticated models even pump warmed water into the wet suits of the divers. This could be useful when working the frigid waters of Alaska. Most divers content themselves by wearing either wet or dry suits to keep relatively warm.

The best dredging is in springtime when the rivers are just dropping in water volume and burgeoning with fresh gold. Dredging season usually runs from the beginning of May into October. A dredging permit is required and a fine levied if one is not in the operator's possession when the authorities drop in for a visit.

In California, the Department of Fish and Game issues dredging permits. Often a small fee is required. Stop in, or phone one of their offices for an application and current fee information for a dredging permit. You must have the permit in your possession when dredging. Be aware also, that size restrictions on dredges may be in effect in certain areas.

Dredging is regulated to protect fish habitats and spawning sites so that dredging does not occur just after fish have laid their eggs. This is a valid concern, as the dredge could suck up the next season's new generation of fish. For this reason dredging is restricted, in most areas, to the months of May to October. The season will vary depending on location and which side of the equator you are on. Generally, the impact that suction dredges have on fish habitats is positive as we shall see in the chapter on "Recreational Gold Prospecting and the Environment."

What most prospectors like about dredging is that there is no digging required. The dredge literally vacuums up the unconsolidated material of the river bed. Most prospectors agree that no digging is a definite plus.

Using a Highbanker To Recover Gold

A highbanker is basically a sluice that rests on metal legs. A hose runs to the river through which water is drawn up to the highbanker via a water pump. Water is released over the sluice bed by an adjustable valve. Gravel is shoveled into a hopper which rests over the head of the sluice bed. A tilted screen, usually made of expanded metal approximately one-half to one-inch mesh, allows gold-bearing gravel to fall through onto the sluice bed for processing. Rocks too large to pass through the screen roll off onto the ground.

Dredging for gold.

Highbankers are used to recover gold from bench deposits, or to process concentrations of gold that may be a distance from the river, or if you abhor the idea of carrying screened buckets of gravel to the sluice. The gasoline pump you select must have enough power to draw sufficient water uphill to the highbanker.

While the carrying of gravel is eliminated with the use of a highbanker, shoveling is required. However, pre-screening your material is not. The highbanker comes with a screen, called a "grizzly," which classifies the material as you dump it into the hopper. Just set up your highbanker where you want to dig and start shoveling auriferous gravel right into it.

Almost any sluice can be converted to serve as a highbanker. Bolt on some metal or wooden legs, and affix a hopper and screen assembly over the sluice head. Attach a simple garden faucet over the head of the sluice so that you can regulate the flow of water over the sluice bed. It is desirable to attach a spray bar to the faucet so that water is directed over the entire width of the top of the sluice bed. Allowing water to simply fall in a stream from the faucet will work, but dirt will begin to pile up on either side of the water stream. More efficient processing of the gravel will occur if you either purchase a spray bar, or fashion one from a garden sprayer or piece of tubing. The rotating spray arms found in automatic dishwashers can be modified to serve as spray bars.

You will also want to be able to shut the water off when the time comes to remove the concentrates for panning, so a garden faucet is perfect. Run a hose from the tubing attached to the faucet down to your gasoline-driven water pump, which is submerged in the river. Fire up the pump motor, and as soon as water begins to pour over the sluice bed, you are ready to start shoveling that gold-bearing gravel!

Some highbankers are manufactured with conversion kits so that you can convert them to a dredge. Merely add the floats, or pontoons, and you have two pieces of equipment for just a little more money, but still a lot less then would be needed to purchase a highbanker and dredge separately. Highbankers can cost between $500 and $1,000 depending on size, manufacturer, type, extras if any, and whether or not they come with a conversion kit. You can build a highbanker utilizing your sluice and save yourself some cash. Or, you can save money by purchasing a used one. Getting into highbanking as a means of finding gold, is generally less expensive then getting into dredging.

Bench deposits, the prime areas worked by highbankers, are composed of the rock, sand and gravel of old river beds. The material composing bench deposits is usually somewhat cemented together, but is still considered unconsolidated material. The boulders composing bench deposits are usually rounded in shape due to the abrasion of rolling against one another as they travel along under water. The further the rocks travel, and grind against one another, the more rounded they become. Rocks composing the benches of desert areas will be more angular due to the intermittent nature of water flow in dry areas, resulting in diminished movement and less abrasion against other rocks.

Bench deposits are one of the richest sources of gold for recreational gold prospectors, and highbankers are how most of us work them. It is a well-known fact that ancient rivers were richer in gold then present-day rivers, due to the fact that nearly two centuries of intensive mining have taken their toll on the gold deposits that, previous to the various gold rushes, had many millennia to accumulate. This does not mean that there are no longer profitable amounts of gold left to recover. There remain many, many overlooked gold deposits. The gold found in bench deposits fits this category nicely, for these are the banks of those ancient rivers. Highbankers are one of the prime pieces of equipment used to efficiently process bench deposits of lucrative gold deposits.

I must issue a word of caution here about working bench deposits. In dredging, there is the possibility of undercutting rocks by suctioning around them. A similar danger is present when working bench deposits. Be careful that you don't undermine large boulders as you dig the material from beneath them. If you find that you are getting close to removing too much supporting material from beneath, or to the sides of, a boulder or group of boulders, you have really only one choice, and that is to create a controlled mini-avalanche of the unstable rock. You must release the loose rock for your protection and that of others.

Use caution and good sense when you do this and stand well to the side. If you miscalculate, you could injure yourself or those who are with you. You could choose to leave things as they are, quit digging, and walk away. However, gravity, rainfall or even the slight vibration caused by aircraft overhead, or someone walking nearby, will likely cause the rock to fall sooner or later. It could be a disaster waiting to happen to some poor soul who just happens to be there at the wrong time.

Rock falls happen spontaneously throughout nature all the time, due to natural processes of erosion, or animal and human movement. Still, you don't want your digging to be the initial cause of a rock fall that might not have occurred for years otherwise.

I frequently have to dislodge insecurely held rocks, some caused by my digging, some that are ready to fall due to natural erosion. I've needed to do it numerous times to prevent rocks from falling out of the bank onto myself or my partners unexpectedly. We always check a section of bench for loose rocks prior to beginning work on it. We usually discuss the best way to accomplish the fall of a loose rock and in what direction we want it to go (sometimes you can control the direction), but most times you can only predict which direction the forces of gravity are most likely to take it.

It is seldom necessary to release more than two or three rocks at any one time. Since bench deposits are fairly well cemented together, usually only your targeted rock or rocks will fall once they are loosened sufficiently. Creating a mini-rock fall does not work on the same principle as an avalanche of mud or snow. Mud and snow are not firmly cemented together. When mud and snow collapse, avalanche, or slide, it usually begins with a minor displacement at the summit. That small portion collapses, then gathers more momentum and additional material as it expands the energy of its own collapse.

In working with bench deposits, generally only the targeted rock, or rocks, will fall out of the bank. Occasionally, a few resting behind will fall out, but it is a very minor amount of material that drops, although the size of one or more of the components may not be minor.

The best policy is to stand well to the side of where you predict the rocks will go. Don't create a rock fall from a portion of the bank over which you are standing. Something could go awry, and the area upon which you are standing could collapse along with the material beneath it.

On one occasion I was prospecting in the Mother Lode region with a friend who had never been prospecting, but had a yen to give it a try. We were working at the base of a bench deposit and processing the gravel with a sluice that was set up about 15 feet away. I noticed that my friend was beginning to undermine a huge boulder about 40 inches in circumference. I warned him to be careful about doing that and suggested that we undermine the rock carefully, due to its size, and release it from the bank, in order to prevent an unplanned rock fall.

After a couple more warnings, I received a rather curt reply to quit worrying about it because he had the situation under control. Irritated at his rebuff, I turned my attention back to dumping a bucket of gravel into the sluice. Later, I realized I should have insisted that we drop the rock, despite the risk of further angering him.

Suddenly, I heard the sound of rock shifting. Without looking back I leapt to one side of where I recalled the last stable position of the boulder to be. Immediately after the first warning shift, I heard the boulder crashing out of the bank. I managed to jump clear of the boulder's path as it rolled swiftly into the river, just missing me and the sluice.

I turned to locate my friend fearing the worst, as he had been working under the boulder. He was very lucky. The boulder had hit him with a glancing blow to the shoulder, knocking him into a tree. His shoulder, knee and leg were skinned up and bleeding, but otherwise he was unharmed. There must be some truth to the old saying, God protects children and fools. Needless to say, I never went prospecting with him again.

Always err on the side of caution and set aside your ego for safety's sake. Furthermore, don't go prospecting with someone who repeatedly refuses to take the necessary precautions. Not only does that person risk his safety, but yours as well, along with anyone else's in the area. Believe me, it is not worth it!

On another occasion I was prospecting with an experienced woman prospector. She was standing in a large hole using her rock pick to loosen gravel to be screened and dumped into the sluice. It appeared that she was working too close to a boulder that wasn't securely set into the side of the hole. I brought it to her attention and she assured me she was not working near the rock but on the other side of the hole. I could see that she was indeed working on the side opposite the rock. Assured, I went back to the hole I was working, which was closer to the river.

After a half-hour or so, I heard the sound of a rock hammer being repeatedly hit against stone. I got up to investigate and when I arrived at her location, I found her with a large rock lodged against her ankle, pinning her to the side of the hole. It took us both to move the heavy rock aside so that she could pull her ankle free. Fortunately, the rock had not broken her ankle when it fell and rolled against it. Still her ankle and foot, despite heavy socks and hiking boots, were severely abraded. Massive, dark bruises began to form as bleeding beneath the skin started when the pressure of the rock against her ankle was released.

Although she had shouted for me to come to her aid, I had been unable to hear her over the sound of the river. Yet the sound of metal on stone, in most cases, will be heard above the sounds of wind, or water, carrying over greater distances than the human voice. Keep this in mind if you ever get into trouble.

This woman was an experienced prospector, yet the unexpected did happen. Although she wasn't digging around the rock that fell on her, she was unlucky enough to be in the way when it did fall. Perhaps it would have fallen at that time even if she were not in the hole. Perhaps the vibration of her digging nearby caused it to fall and roll up against her ankle. Whatever the cause, she maintained her cool, and found an effective way to let me know that she was in trouble.

These two examples are given so that you will fully realize the dangers of working around large, unstable boulders. If you operate with caution and foresight you can eliminate a lot of potential danger to yourself.

In nearly 20 years of gold prospecting I have never been injured by rock that I have been digging around. Perhaps the main reason is that I am cautious. I've had a few close calls by being in the same area in which others were digging who did not take the proper precautions or maintain awareness of what they were doing, let alone keep in mind the safety of those around them. Experience has taught me that over-caution concerning large boulders is a good thing to develop.

If you decide that you must avalanche a few rocks from the bank for safety purposes, make sure that no one is in the way of the rock fall. Let anyone in the immediate area know what you are going to do. This way no one inadvertently steps into the path of the falling rock. Maintain your awareness of your surroundings and you will have a lifetime of safe, enjoyable — and profitable — trips with your friends and family.

Drywashing for Gold

Drywashing for gold is a method by which gold is extracted from gravel without the use of water. Drywashers depend upon air and vibration to separate gold from dry desert gravel. A drywasher uses a riffle bed similar to that of a sluice. The riffle bed of a drywasher, instead of the solid wood or metal bed of the sluice, usually has a fabric bed which allows air to be forced up between the riffles. Gold lodges behind the riffles in a "dead space" located behind each riffle. This dead space is usually composed of a metal or wooden bar which prevents air from dislodging any gold settling there. Air is provided by a bellows that is operated by a gasoline engine, or in some cases a battery-operated motor. The bellows forces air up through the fabric of the riffle bed in a rapid series of bursts or "puffs."

*The author
digs down
to bedrock
in a desert
wash.*

The combination of air and the vibration caused by the operation of the drywasher separates out the gold and heavier materials. The riffle bed is set on an angle, which is adjustable by the operator. Dirt shoveled into the hopper and through a grizzly, also set on an angle, falls upon the riffle bed and is classified as it is shaken along. The grizzly separates out larger stones that fall into a pile.

To drywash for gold you need:

1. A drywasher.
2. Shovel.
3. Hoe pick.
4. Gold pan.
5. A gallon of water.
6. A small tub or other receptacle for panning into.
7. A bandanna.

The angle at which you set your riffle bed determines how often you will have to clean out your concentrates and the efficiency of the drywasher in retaining gold. Experience with the drywasher and some testing with metal fragments will resolve the angle at which your drywasher is most efficient at capturing gold. You will want to set the angle for maximum gold retention and yet maintain the efficient reduction of the lighter, unwanted materials. Too slight an angle will cause the riffles to fill too quickly and require you to stop the drywasher and empty it of concentrates. Also, the concentrates will contain too much useless, lighter material, which you will then have to pan.

Drywashers are used to process the auriferous gravel of desert washes and alluvial areas. Gold may travel a long way from its source in the desert. Over a series of decades, flash floods and runoff from rare rainstorms cause gold to travel ever further from its source.

The author pours water into an 18-inch pan. She will pan into it with a 10-inch pan.

Gold settles out and concentrates in dry washes much the same way as it does along and within rivers, creeks, and streams. The problem with desert prospecting is always a lack of water. You must bring enough water to supply your needs and your prospecting requirements.

I bring ample drinking and cooking water for myself, and extra for the vehicle in case a radiator hose breaks. I bring a couple of gallons to be used for panning my drywasher concentrates. Two gallons is sufficient for a prospecting trip lasting several days.

I use a large 18-inch metal gold pan as my panning receptacle. It holds about one quart of water. Then I use a smaller 10-inch pan to pan with. A plastic tub can also be used to pan into as long as you have ample room and depth to manipulate your gold pan. Although the water in your panning tub will become discolored with dirt, it will not detrimentally affect the panning process or recovery of gold. When the panning water becomes thickened, it is time to dump it and the silt and sand lying at the bottom, and pour in fresh water. Water from washing, shaving, and dish washing (as long as there is not any grease) can be recycled for use as panning water.

The author shovels dirt into a drywasher working on a desert site. Note the bandanna for protection from dust inhalation.

One of the problems you may encounter in desert prospecting is the dryness of the gold, which sometimes prevents the fines from breaking the surface tension of water in your gold pan. You will see these minute yellow flakes swirling about on top of the water as their specific gravity battles the surface tension of the water. This is easy to remedy by squirting a small amount of liquid dish-washing detergent into your panning tub. You don't need much; the point is not to create any suds. Just a few drops is sufficient to reduce the surface tension enough for gold to sink. That recycled wash water, mentioned previously, fits the bill perfectly!

Drywashing can produce prodigious amounts of dust, depending on the fineness and dryness of the dirt in your location. Wind can improve the situation, or make it worse, depending on which direction it is taking the dust. You may notice that many desert prospectors wear bandannas around their necks or hat brims. This is not a fashion statement peculiar to desert prospectors but a protection from dust. In areas where the dust is particularly fine, the bandanna is tied around the head — just as an old-time bank robber would wear it — so that it covers the nose and mouth, allowing the operator to breath without inhaling any dust.

Drywashers are usually operated by one or two persons. One may shovel while the other uses the hoe pick to loosen earth for the shovel operator. In areas where the wash benches and gravel are not cemented, both operators may shovel dirt simultaneously into the drywasher.

While drywashers process gravel with air and vibration, electrostatic concentrators use the additional resources of heat and a small electric charge. The heat slightly dries any material that is damp and the static charge allows gold to cling better to the riffle bed. Electrostatic concentrators usually cost a little more than drywashers but work on the same principle while allowing operators to work in areas where the soil is slightly (notice I said "slightly") damp. Drywashers will not process damp gravel of gold efficiently.

You can work slightly damp gravel with a drywasher by digging out the dirt that you want to process and spreading it out on a tarpaulin to dry in the sun for a time. You can also process the damp gravel once through the drywasher. The damp gravel is subjected to separation and exposed to air as it passes over the riffle bed. In some cases, where the gravel is only very slightly damp, this is enough to dry it sufficiently for a second trip through the drywasher. Simply shovel up the gravel after its first pass through and dump it back into the hopper for a second trip. If this fails to dry the material sufficiently, spread it out on a tarp as suggested previously. If the material in the area you are working cannot be dried by any of these means, then you will have to come back in several weeks, or months, after warmer, dryer weather has dried the material.

You can purchase motorized or manually operated drywashers. Using a manually operated drywasher means that the operator will be doubly tired at day's end. If you decide to purchase one, look for a model that is crank-operated, not the "rope-pull" type. Depending on features, expect to pay from $500 to $1,000 for a top-of-the-line model. Once again, you can purchase kits for making drywashers

Spreading damp soil with a rake to dry it before drywashing will maximize gold yields from this wash method.

or you can purchase a used one. You will save yourself quite a bit of money this way. A lightweight, manually powered drywasher, suitable for backpacking, will cost around $200.

Something to keep in mind when drywashing for gold is that you may encounter nuggets that are too big to pass through the grizzly. The grizzly is usually composed of one-quarter-inch to one-half-inch screen mesh. Nuggets bigger than this will not pass through the grizzly, but will pass over the top of it and into your detritus pile. Always, check your rock pile for any nuggets that may have passed over the grizzly. The most efficient way to do this is with the use of a metal detector. Pass the detector over the rock pile. Then sweep off several inches of rock and detect the next level. If you do not have a detector, you will have to make a visual check through the rock pile. You must run any dirt at the bottom of the pile through the drywasher **with the grizzly removed** (you don't want to screen out those big pieces of gold again). Gold will naturally seek the lowest point of your pile as you disrupt the various layers of rock in your quest for those possible lost nuggets.

This is a good place to mention the age-old art of dry panning. Dry panning is a form of panning that is done without water. It too is dusty work, so have your bandanna handy. To dry pan you merely place the dirt you want to sample into your gold pan. Then toss the contents into the air, catching them as then come back down. In this way the wind actually winnows away the lighter materials, and gold — due to its specific gravity — falls back into the pan, along with black sand and rocks, if you haven't removed them. It is a good idea to remove rocks and most of the gravel prior to dry panning. This method is not very efficient, but it will work in a pinch. I used it with success many years ago on a desert prospecting trip. Needless to say, I don't prefer it, but it did come in handy.

In some areas the use of any motorized recovery equipment without a permit may be subject to penalty. Recent legislation requires that any motorized equipment — be it dredge, highbanker, drywasher, or lawnmower — requires a permit in some areas. It is always advisable to contact the agency of jurisdiction over your site to check on regulations regarding your equipment before you go there.

Besides metal detecting, drywashing is the best way to recover gold from dry desert areas. Drywashers are used to recovery fine gold and the nuggets that are deeply buried beyond the range of metal detectors. A winning team is created when you take both a drywasher and a metal detector into the desert. Also, fewer prospectors are desert prospectors, which means that profitable amounts of gold are waiting for you!

Dredges, highbankers, and drywasher concentrates are panned in the same way as sluice concentrates, with only minor variations. Drywasher concentrates do not need to be rinsed into an empty bucket. They are dropped in dry for subsequent panning. The basics that you learned earlier will serve you over and over when you utilize other types of recovery methods. All of these other methods are merely variations on an original theme — a theme that you learned with your gold pan and sluice!

Electronic Gold Prospecting

Electronic prospecting — gold prospecting with a metal detector — is one of the easiest and most profitable ways of finding gold, as well as other metallic treasures. A devotee of both sluicing and drywashing for gold, I find that electronic prospecting, or "nugget shooting" as the insiders call it, is gradually supplanting my other methods. The reasons are many, the main one being the quantity and size of nuggets I get with only minimal effort. Too, most other methods require packing along quite a bit of stuff, unloading the stuff, carrying the stuff to the site, setting the stuff up, then, when its time to leave, going through the entire "stuff transporting" process in reverse.

I find it so easy to just grab the detector, and the small gym bag that accompanies it, and go! Am I getting lazy? No, I still appreciate the other methods of gold prospecting and enjoy them, but I have grown to love the ease and success of electronic prospecting.

I can wield a detector all day long and not be tired. The most exertion I undergo is to occasionally bend down to dig a target. Often the gold I find with the detector is quite abundant because I can hunt areas where the ground is too wet for drywashers, or where there isn't enough water for even rockers — let alone sluices or highbankers. If water is present, it's not always deep enough to support a dredge. The nuggets I get with my detector tend to be larger than those recovered with more traditional methods because I am able to hunt areas that cannot be hunted with most other types of gold recovery equipment. With my metal detector I can hunt anywhere, anytime.

It has been asserted that dredging is one of the most popular methods of recreational prospecting. This may be so, but I suspect that before long — if it has not already occurred — electronic prospecting will soon surpass dredging in popularity. Too, a high-quality gold detector can be purchased for considerably less than most dredges. Another positive aspect of nugget shooting is that your detector is your sampling tool, and except for digging your target for you, may also be considered your gold recovery tool.

Traditionally, gold prospecting has been the activity of men, except for a few unconventional women who have found that nothing compares to the fun and adventure of recreational gold prospecting. With the advent of gold-finding detectors, I feel that more and more women will become prospectors. Electronic prospecting is likely to be the introduction to prospecting that numerous women have been waiting for.

I am often asked, "Why aren't more women gold prospectors?" I suspect a variety of factors. Partly, advertising may be to blame. Most of the advertising is aimed at men. If women are depicted in advertisements, it is usually for the purpose of attracting the eyes of men to the advertisement. An attractive woman may be used to visually embellish a piece of equipment — not as a potential user of the equipment, but as a prop. This type of advertising may work well with automobiles, but I think it is lost on most serious gold prospectors. Furthermore, while this type of advertising may attract the eyes of novice male prospectors, pros tend to have eyes only for the equipment. Anything else in the photo is just a distraction — except, perhaps, a nice chunk of gold!

A small but growing number of women, myself included, enjoy the exercise and the challenge of finding gold in traditional ways, but I suspect that most women just don't see why they should engage in the transport and operation of large pieces of recovery equipment (sluices being an exception). I think many women are just turned off by the whole idea of shoveling dirt! Under most other circumstances, so am I, but for gold — well, let's just say that I have an incurable case of Gold Fever and shoveling dirt is not a problem. It's a means to an end — a profitable end!

Add to this the fact that many women have been socialized to be observers of the activities that men are engaged in, rather than participators. They accompany their spouses and boyfriends on prospecting trips, but remain in the wings socializing with one another, while the men are getting all the gold.

The reason that more women are not gold prospectors is merely perception. Many men, and most women, just don't perceive their roles and possibilities as different from how they were conditioned by family and society as they grew up. This is not so true of young girls today, but women of my generation, and older, were more constrained by attitudes regarding sexual stereotypes. I, the eternal rebel, was just never influenced much by the expectations of others when it came to something that I wanted to try that looked like a whole lot of fun in the bargain! On this subject, I would like to announce, "Times are a' changin'!" The gals are getting tired of sitting on the sidelines letting the men have all the fun, get all the glory — and all the gold!

Too, women may feel disadvantaged by the fact that men are generally stronger than women. Perhaps they feel that they could never hope to compete with the guys on this level. To this I say, "Fine!" But allow me to add that a successful prospector is not determined by strength. The two factors which do determine a successful prospector are **research** and **sampling**. No strength needed there, just patience — something women have been cultivating for eons!

A woman **can** move just as much dirt as a man. It just may take a little longer. Most women have endurance rather than strength. As an example, when onto good gold-bearing dirt, I may be moving less dirt, in a given period of time, than my male counterpart, but I have the endurance to work for longer hours. So, it all balances out in the end. Nature gifted men with more physical strength, but gifted women with greater endurance — ask any woman who's gone through childbirth. Then ask her how her husband would have fared had it been him instead of her!

Once bitten by the gold bug, most women will be right there in the mix, sampling, shoveling, panning — whatever it takes. Once a woman finds gold in her pan — by her own hand — she will not be satisfied with sitting on the sidelines.

Another factor on the side of women is intuition. Now, I believe that both men and women were gifted by nature with intuition. Yet again, the expectations of society have done men a disservice by subtly, and not so subtly, indicating that intuition is a feminine virtue not to be claimed or given much credence by men. It just isn't considered macho to "have a feeling." Men are a little more at ease with the concept of "intuition" if they can get away with calling it a "hunch." Call it intuition or a hunch, it is an edge invaluable to the prospector — man woman, or child — that is worth cultivating. Due to the fact that women don't feel constrained to stifle their intuition, this is an edge that women can bring successfully to their gold prospecting, an edge that more men should be keener to cultivate.

Nugget shooting is a method of gold prospecting that I believe women could get into in a big way. It's not hard work. No great strength is required. Patience is useful — something most women have in abundance. A metal detector is light in weight, and unlike most other types of gold recovery equipment, does not require a truck to transport it to the hunt site.

Electronic prospecting is also a great method for those who do not feel that they have the energy or stamina for other methods of prospecting. Those who are elderly or physically handicapped can enjoy recreational gold prospecting with a metal detector.

Electronic gold prospecting is also great for those who are full-time RVers. A metal detector does not require much in the way of storage space and is light in weight — a perfect combination for the RVer or apartment dweller. I live in a condominium. My single-car garage is crammed — from top to bottom — full of prospecting gear, lapidary (rock cutting/polishing) equipment, and my Suzuki Samurai expedition vehicle. On the other hand, my three Gold Bug detectors occupy a narrow spot in the closet of my computer room.

There are detectors on the market for all purposes, such as coin hunting — called "coin shooting" by enthusiasts — treasure hunting, treasure diving, and gold prospecting. The type of detectors we are interested in here, as recreational gold prospectors, are those that are manufactured specifically for finding gold. These gold-finding detectors are the "greyhounds" of all the types of detectors available today. They are sleek, specialized in function, and extra-lightweight.

Gold-finding detectors have extraordinary sensitivity and depth.

This dandy nugget was recovered with a gold-finding detector.

Gold is one of the most difficult metallic substances to detect, so a high-quality detector — specifically manufactured to locate gold — is what you want for recreational gold prospecting. Sure, any type of detector will find a large, not-too-deeply-buried nugget, but it's those small, deeply-buried nuggets that are the prospector's bread and butter. For this you need a special type of detector that can pinpoint those nuggets and at the same time be able to negotiate its way around ground mineralization.

The author, enjoying a beach hunt, prepares to dig for a coin targeted by her gold-finding detector.

If you want to spend an occasional day coin hunting at the beach with the family, you can use your gold detector to do that, too. I do it all the time. I collect my niece and nephew, take my three Gold Bugs — one for each of us — and we spend the morning on the beach finding coins, wading in the surf, and just having a fun outing together. On lucky days we may find an occasional ring, watch, or other metallic treasure, along with our coins. Afterward we like to treat ourselves to ice cream or some other sweet treat — often paid for with the coins we've found!

Occasionally I use my detector to hunt for caches and coins around old mining settlements or look for interesting relics around mines that I encounter as part of my prospecting trips. But first and foremost, I use my detector for nugget shooting!

Recreational gold prospectors owe a huge debt of gratitude to Dr. Gerhard Fisher, who studied electronics at the University of Dresden. He immigrated to the U.S. in 1923. In 1931 Dr. Fisher, by then living in Palo Alto, California, turned his backyard garage into a research laboratory, which he named the "Fisher Research Laboratory." In that small garage he began inventing some marvelous things. In the 1930s Dr. Fisher was best known for the development of the radio direction finder, which soon earned him a contract for their manufacture in airships, better known as "dirigibles."

This application of electronics attracted the attention of the noted physicist, Dr. Albert Einstein who paid a visit to the young German inventor and was inspired to predict, "Its use (the direction finder) for navigational purposes would, within a few short years, be used as required equipment throughout the entire world, in the air, on the land and at sea."

Would that Dr. Einstein had been so positively inclined toward another of Dr. Fisher's patents — the "Metallascope." Of the Metallascope, Dr. Einstein stated to Dr. Fisher that he "didn't think the Metallascope was a very useful device." Seldom has history — or anyone, for that matter — proven the great Dr. Albert Einstein wrong — except for Dr. Gerhard Fisher, inventor of the Metallascope!

The Metallascope proved to be the first successfully manufactured and patented (1933) metal detector in the world. Within a few years of its development the M-Scope went into production at the Fisher Research Laboratory.

Gerhard Fisher (right) with Albert Einstein in the 1930s.

Before long the potential of the M-Scope became well-known. It was used in police work, as early as 1937, to recover lost murder weapons. It was successfully used by undertakers to locate long-buried coffins, by industries and homeowners to locate pipes and phone lines. Also in 1937 a miner at Jackson Hill, in the Mother Lode region of California, used an M-Scope to locate a pocket of gold worth $1,500. It was not long before other miners, treasure hunters, and hobbyists discovered the value of a metal detector.

Some of Dr. Fisher's employees eventually went into business for themselves to undertake the manufacture of metal detectors under their own names. Inspired by Dr. Fisher's successful patent, there are a variety of detector makes and models presently on the market.

Today, the popularity of nugget shooting is enormous and still growing. What started out as the Metallascope, or M-Scope, has become for gold prospectors, treasure hunters, and coin shooters, the machine of choice for the hunt. In fact, Dr. Fisher's original company, the Fisher Research Laboratory, still manufactures high-quality metal detectors for industrial use, as well as for both hobby and professional gold prospecting, treasure hunting, and coin shooting. In honor of Dr. Fisher, the company's detectors still carry the name "M-Scope."

My detector has proven invaluable when nugget shooting in both mountains and deserts, and over the years I have developed some techniques which have led to success and profit. It is my hope that you will profit from what years of nugget shooting have taught me.

I always make it my habit to pass the detector carefully over piles of drywasher tailings that I may encounter on my trips. Many prospectors who use drywashers do not take the precaution of detecting their own drywash tailings. Perhaps they don't have a detector along.

For a variety of reasons nuggets may pass over the riffle bed of a drywasher and land in the tailings pile. Perhaps the dirt might be too damp or the riffle bed set at too steep an angle. Occasionally a nugget, or nuggets, can be gleaned from someone's tailings pile. I have found nuggets by doing this, and chances are good that you will too.

When I am drywashing, I always take along my detector, not only to detect my tailings and the tailings left by others, but also to do a bit of nugget shooting when I want a break from shoveling dirt into the drywasher.

Dry desert washes are great places to find nuggets, and so are the areas around outcroppings of rock, which are quite commonly found in desert areas. Tailing piles at mines, and the areas around mill sites, are also spots that you want to hit with the detector. More than one prospector has pulled a gold-laden chunk of ore from a mine's tailings pile that his detector signaled was buried there.

The foundations around miners' cabins are spots to keep in mind for potential gold caches. Most miners had a habit of burying their "pokes" (small caches of gold) around the foundations, or in the walls of their cabins. Access to banks was not always handy, as it is today. In remote areas, the safekeeping of valuables was usually accomplished by burying them.

The best way to find gold with a detector is to do your research and prospect in areas where gold has been found before. Don't forget that old-time prospectors believed that the presence of outcroppings was a possible indicator of gold, and there is valid geological background for this assumption, having to do with the decomposition of vein material and subsequent zones of mineral enrichment.

Using a detector to prospect for desert gold eliminates the need for any water at all, other than what you tote along to quench your own thirst and, of course, to supply your vehicle should you break a radiator hose. Most detectors will not pick up fines, but are dynamite for finding both large and small nuggets. The detector, of course, eliminates the need for sampling.

Old mine cabins are great places for "cache" hunting. Many miners mistrusted banks and buried their gold.

All detectors come with an instruction manual that gives tips for tuning them, along with hints on how to "target" a nugget and to re-

Relic hunting in mine camps might yield cutlery, rail spikes, medicine bottles, and cupels for pouring buttons of refined gold.

trieve it. We will go over the basics of electronic prospecting, but for tuning your detector you will have to follow the specific instructions that accompany the particular detector you are using. There will be some variation, depending on manufacturer. In general, here are some things you will want to look for in a gold-finding detector.

Detecting near the ruins of a mining cabin in the southern California desert.

1. Make sure the detector you purchase is manufactured specifically for nugget shooting.

2. An elliptical, rather than round, search coil will make detecting between rocks and around bushes easier.

3. Expect to pay $500 or more for a good detector, $700 to $800 for a great one.

4. If the detector you are considering for purchase has too many complex instructions for tuning, it is better to forget that model. If you cannot understand what is required to tune the detector, it will not be satisfying to hunt with it.

5. Look for one that has a "ground reject" or "ground balancing" feature for adjusting to varying ground mineralization in the area where you are hunting.

6. You also want to be able to control the sensitivity settings, so look for a detector that allows you to do this.

7. Look for a model that is light in weight and has a removable control housing that you can hang on your belt. After a couple of hours, a detector's weight will make the difference between whether or not you want to continue.

8. Be sure that any detector you use has a plug-in for earphones. Earphones are a must to screen out background sounds of wind, water, and traffic sounds. Earphones allow you to hear the faint signals of deeply buried nuggets.

9. Make sure your detector has a "battery test" feature. A quick test of your batteries will save you a lot of guessing.

10. Make sure the search coil is waterproof — a must for nugget shooting shallow water areas!

11. Be sure that your detector has the capacity to reject "hot rocks" (more on this later). This feature is a must!

12. Avoid detectors that use a lot of batteries. Batteries add weight.

13. Spend a couple more dollars and buy a "snap-on detector coil protector." You could use duct tape to protect the coil bottom from abrasion, but it looks tacky and detracts from the clean lines of your sleek new detector.

The method used to detect for gold is to swing the detector back and forth in front of you in a smooth arc, almost making a half circle in front and extending to either side. Make sure that the detector coil just skims the ground, keeping the coil parallel at all times to the ground. Be careful that it doesn't lift from the ground when it reaches the apex of the swing at each side.

If you have a large area to cover, walk in one direction, then turn and come back the other way so that your sweeps with the detector overlap. This way you cover the ground thoroughly. To nugget shoot you will need:

1. A good gold-finding detector.

2. A small shovel, or hand trowel.

3. A rock pick.

4. A small plastic gold pan (not for panning — did I mention the fact that you do not have to know how to pan to prospect with a detector?).

5. An empty plastic film canister for the gold you find.

6. A belt that will hold your rock pick, plastic pan, trowel, and possibly your metal detector control housing.

I wear a leather belt, not to hold my trousers up, but upon which to hang my rock pick and trowel. When purchasing a rock pick, it is also possible to buy a holster for it that will slip onto your belt. Holsters come in metal, leather, or a combination of the two. The pick can be easily slipped in and out of the holster for use, leaving your hands free. The trowel that I use is a called a "U-Dig-It" (available mail-order from Contract Geological Services and many prospecting supply stores. See the chapter, "Outfitting the Prospector"). It is forged of sturdy stainless steel and folds into its own nylon carrying sheath, which slides onto my belt. My plastic gold pan has been drilled with a hole so that I can hang it on my belt. The pan hangs from a metal shower curtain clasp that hooks over my belt.

A removable detector housing can be slipped onto your belt in a couple ways. Most housings come with a built-in belt loop. Or, you can purchase a fabric holster for the housing which then slips onto the belt. The fabric holster is nice because it helps to protect the housing from dirt and scratches.

I usually don't bother with detaching the housing because the detector I use is so light weight (only three pounds) that I find I can hunt all day with the housing in place. But it is good to know that I do have this option if I ever want or need it.

The main problem encountered when nugget shooting is that gold is usually found in highly mineralized areas. Gold is one of the hardest metals to detect and this is where the problem arises. Iron is one of the easiest metals to detect. So the problem is to create a detector that rejects mineralization (most of which consists of various iron compounds) while remaining sensitive to gold. This has required an entire separate technology, which is why coin shooting detectors are not good for nugget shooting. Most coins are not found in highly mineralized soil, but in parks, at beaches or at campgrounds where soil mineralization is essentially neutral. Surmounting the problems of mineralization is not a factor in manufacturing coin-finding detectors.

Because gold is the most difficult metal to detect, gold-finding detectors are dynamite on coins that are made from easier-to-detect metals. You can use a gold detector for the occasional coin hunt! If you really get into coin shooting you will eventually want to invest in a good coin-finding detector. They have a feature called "discrimination" which screens out common junk like nails, foil, pull-tabs, bottle caps, and the like. For nugget shooting you really don't want too much in the way of discrimination (except for hot rocks) because you could end up discriminating out the gold signals.

Technology has been able to solve much of the mineralization-versus-gold dilemma, to the delight of recreational prospectors. The problem that is still being faced by most manufacturers of detectors is that of "hot rocks." Most have solved this problem somewhat, but some have solved it better than others. Therefore, I have a couple of suggestions concerning purchasing a detector.

While it is not my place to recommend one manufacturer over another in this book, for all put forth a good product, some products are better than others. I would suggest that you observe other prospectors in the field. See what they are using. Note particularly the brands that are used by the pros for nugget shooting. In this case a picture is worth a thousand words!

Talk to prospectors you encounter in the field. Most love to talk about their detectors (if they are happy with them). I know I love to talk about mine. While a prospector may not tell you if they're any getting gold (some will even be evasive about it), most will be willing to discuss their detectors. If they offer to sell it to you at what seems a ridiculously low price, be suspicious. They may not have done their homework and ended up with a detector that doesn't meet their expectations.

You might try renting several brands of detectors before you commit your hard-earned dollars to one. Many shops that sell equipment also rent detectors. You may spend some money trying the various brands, but when you go to commit your dollars, you will be assured of getting a detector that you know will be a success in the field, with a minimum of fuss, for many satisfying years to come.

Hot rocks are nodules, mostly of iron. They look, in many cases, like innocent rocks. They are not! If your detector sounds on what looks like an ordinary rock, this is your first clue. Pick the rock up. It will feel heavier than normal — your second clue. In many cases it will look rusted in a few spots — your third clue. This is a hot rock and it is the bane and nemesis of the nugget shooter. And they are usually all over the place where gold is found. This is the reason that you must get a good detector that will reject hot rocks. Otherwise you will be driven to near insanity, if not abject frustration.

A couple years ago I was detecting, with a friend, in an area of northern California that had once been subject to hydraulic mining (a process where tremendous volumes of water are used to tear apart gold-bearing sediments. Don't even think about it; it's now illegal!). This area, while rich in gold, has two problems. One was electric power lines. It was nearly impossible to detect under power lines without registering audible interference from the lines through the earphones of our detectors. The other problem was numerous marble-sized hot rocks. Realizing that the electric lines were an obstacle, I still hadn't a notion that hot rocks were a larger one. I had encountered two or three pesky ones, which I immediately rejected by switching to my "iron discrimination" mode, but wasn't overly aware of a big problem with them as my detector rejects the majority of them. My friend however, was nearly tearing our his hair. Every swing of his detector caused it to sound off on buried hot rocks and so he had to dig a lot of bogus signals.

He was using a different make of detector from mine, and that trip really pointed up the differences to be found in the gold-finding detector market. So be sure that you are getting one that rejects the majority of hot rocks. Most manufacturers make the claim that theirs rejects hot rocks, and all do to a point. Again, some do it better than others. Believe me, you will be glad that you did homework on this one. By the way, we found some really dandy nuggets on that trip. My friend got his nuggets, but risked the loss of some of his hair!

Gold-finding detectors will work just fine without earphones. Earphones are worn to screen out some of the distracting noises of wind, traffic, and water. They also help convey directly to your ears the faint sounds of deeply-buried nuggets. In normal operation, the speaker on the detector control panel is about three feet from your ear. It is possible that you might not hear faint signals unless they are conveyed directly to your ears by earphones. The earphones don't block out all sound. You will still hear birds and other noises. The earphones merely allow you to hear detector signals as a "priority sound."

Earphones will also help to preserve the life of your batteries. When using earphones you can keep the volume control at a greatly reduced level, which extends the life of the batteries. Just ask any kid with a "boom box" who goes through batteries lickety-split, just because he likes it loud!

Remember that what is a comfortable sound level setting without earphones will be an intolerable level with earphones. So when you first turn on your detector and you're wearing earphones, don't crank up the sound at the same time (the on-off switch and volume control are usually the same knob). Just turn the knob until it clicks on. Then gradually increase the volume to a comfortable level.

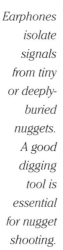

Earphones isolate signals from tiny or deeply-buried nuggets. A good digging tool is essential for nugget shooting.

Once your detector is turned on you will have to "ground balance" it. This procedure varies with different makes and models. Some are easier to tune than others. A friend of mine has a very exotic and expensive detector that she uses for prospecting. It is very complicated to tune and she is never sure that it is tuned properly. It probably never was because mine, which is easy to tune, detects circles around hers. So be sure that any detector you purchase is not too complicated to tune or you too will never be certain that you're not passing over nuggets, but not receiving signals.

You must tune your detector when you arrive at the area of your hunt in order for it to properly reject ground mineralization. Don't tune it to the Berber carpeting on your living room floor, then expect to travel several hundred miles away to hunt gold in an area with a really radical geological history and varied mineral palette. It won't work. Don't tune your detector until you are at your hunt site. If you drive to a new hunt site, retune your detector. It only takes a minute.

Let's say that you've done your research and found a likely spot for gold. You have your detector properly tuned, your tools hanging on your belt, your earphones on and plugged in, and an empty film canister in your pocket for those gold nuggets you're going to find. You're ready to hunt! What comes next?

Reading the river! Or, in the case of desert prospecting, reading the wash. The same rules for finding gold still apply. Concentrate your search on the types of places where gold settles due to its specific gravity and the obstacles present that may slow the flow of water — and, of course, on areas where gold has been found historically.

A desert wash may be harder to read, as will be other areas of intermittent water flow. You can look at a river and project what it might have looked like at flood because you are having the experience of what it looks like now — a starting point, if you will, whose present water flow acts as an indicator from which to estimate, or project. A dry wash is another proposition. Did water rage through the wash, or just flow slowly? You may find evidence in the form of disturbed vegetation. However, many desert areas have so little rainfall from year to year that vegetation can grow back, leaving few clues. Utilize the basics and prospect for all eventualities of water flow.

Speaking of vegetation, there is a type of prospecting called "biological" prospecting. This is an entire subject for study in itself so I won't get into it in any detail here except to say that certain types of plants grow in various types of soil. Some plants prefer mineralized soils, such as the soil types where gold occurs. By studying vegetation, it is possible to determine likely spots where minerals may occur. This holds true for gold as well. The U.S. Geological Survey has some information pamphlets on this subject; information can often be found at prospecting conventions.

Here are just some of the types of places to prospect with your detector:

1. Desert washes.
2. Placer and alluvial areas.
3. Around outcroppings.
4. Mine dumps and tailings piles.
5. Mine mill sites.
6. Along rivers, creeks, and streams.
7. Shallow underwater areas of rivers, creeks, and streams (the detector coil is waterproof — if it's not, don't buy it). Don't ever get the control housing wet.
8. All of the places where you would sample pan.
9. Around abandoned miner's cabins.
10. Around areas where lode veins are found.
11. Drywasher tailings.
12. Old hydraulic mining areas.

13. Tailings from old bucket-line dredges.

14. Bench deposits.

15. Cracks and crevices in bedrock.

There are other types of places too, but these are all good bets. Here is a hint that many recreational prospectors do not know: Gold is often found parallel to deposits of serpentine. Serpentine is a fairly common rock that is seen in road cuts, particularly in northern California's Mother Lode, in some places in southern California, as well as in many other parts of the world. It occurs in various shades of green, yellowish green, white, brown, and black. It has a silky or waxy luster. Keep this one in mind! If you don't know what serpentine looks like, visit the geology department of a nearby college, the mineral section of a natural history museum, or a rock shop.

As mentioned above you can detect shallow water areas. Just DO NOT get the detector housing wet. If you do you will probably ruin the detector. You can detect in a light rain or drizzle by carefully wrapping the housing in plastic. I don't recommend detecting in heavy rain as some moisture is bound to find its way into the delicate electronics. Also, it is not safe to prospect in the rain. Rocks become more slippery, lightning strikes are a danger, and rivers can rise quickly when fed with water running off mountain slopes. Go back to camp and have a cup of hot chocolate!

A metal detector is a great aid to crevicing. I like to sweep the area that I want to crevice and note which crevices I get a signal on. I dig those first and pan out the contents (crevicing is one form of prospecting with a detector where panning is useful, although not essential). At my leisure I will dig out the other crevices for fines and flour gold.

Say you're prospecting in the desert, or along a river, and you get a signal. Pinpoint your target with the detector coil by passing it back and forth in one direction then in the other. The point at which you get the strongest signal will indicate the area below the coil where the target is located. My detector has a little bull's-eye decal on top of the coil. Usually the nugget is located right under this area when I get my strongest signal. Note the spot. Using your trowel, dig right there. If the ground is hard, use your pick to loosen it. Dig some scoops of dirt and drop them into your plastic pan. Pass the detector coil over your hole. If you still get a signal, the target is still in the hole. Dig another scoop into the pan. When no signal is received by passing the coil over the hole, the target should be in the pan. Pass the coil over the pan to verify this. Now you see why you must use a **plastic** pan when nugget shooting; otherwise, the detector will register the metal pan and not the nugget. Be aware that it is possible to have more than one target in a hole. Nuggets sometimes group together. So, even if you get a signal that a target is in your plastic pan, always pass the detector back over the hole so you don't leave any nuggets behind.

Shaking the target-containing soil over the top of the detector coil isolates the target.

The easiest way to separate, or isolate, your target from the dirt in the pan is to shake (using a gentle, measured shaking motion) the pan contents onto the top (yes the top) of the detector coil. Be sure that you are not wearing any rings on the fingers, or a watch on the wrist of the hand that wields the pan; the detector is very sensitive and will sound on those items. Be sure your digging tools are beside or slightly behind you (or on your belt), or your detector will sound on those, too. If you have metal eyelets in your boots, as I do, be sure your feet are not in proximity to the detector.

You can when your target (hopefully a nugget) is nearing the edge of the pan because you will begin to get a "jiggling" sound from the detector. When you have a little pile of dirt on the coil top, dump it off. If the target lands on top of the coil you will hear the signal it makes, kind of a "zing-bang." Dump the dirt and target from on top of the coil back into the pan and continue sorting in this way until very little dirt is left in the pan. You should be able to spot your nugget in the bit of remaining dirt in your pan. Put the nugget into your film canister.

I use film canisters when nugget shooting because quite often the nuggets I find are too large to fit into glass vials. Too, I can put a plastic canister into my pocket and not worry about it breaking when I stoop down.

After becoming familiar with the sounds that your detector makes, you will begin to understand its language. You will begin to distinguish the sounds that different metallic substances make. For instance, nails and wire create a double signal from my detector. But still check the target just to be sure. Before long you will understand the particular tonal quality your detector makes when it is passed over gold. Your pulse will begin to pound when you hear this tone or signal. Even so, I still dig all signals because conditions can serve to alter the sounds that gold may make. A big nugget can sound different from a small one under some conditions. When I hear a certain high-pitched scream, I know that I am onto gold!

There is another type of prospecting that can be done with a gold-finding detector – the finding of "pocket" gold. When most prospectors refer to "gold pockets," they are talking about gold that is found in concentration — a little "glory hole." Gold pockets often contain gold from a variety of sources that has been concentrated into a spot, or pocket, due to water movement. Gold pockets and pocket gold — two similar sounding terms that mean something quite different!

Pocket gold is another affair altogether. While it may also be found in pockets, it refers to gold that has weathered out of its source vein and not traveled too far from the original site. It is usually rough and angular, not rounded and smoothed like nuggets that have traveled a distance from their source and are found elsewhere. The gold referred to as pocket gold is also found in concentration, but its concentration originates from a single deposit or lode, not by accumulation from several sources.

Before the advent of metal detectors, pocket gold was located by sample panning and following the pattern of concentrates back to the original source — usually on a hillside. This was very time-consuming and required stakes to mark the pattern of concentration in order to eventually follow the pattern — usually somewhat triangular — back to the origination point. It could take days and much panning to track it back to its source.

Usually the prospector had to tote along panning water as he moved further up the hill in order to pan the samples that were dug. Dry panning was an option if the soil was not damp. I have done this type of prospecting and it is a complex and work-intensive task. Sometimes it is worth the effort, sometimes not. If the gold is fine or flour gold you must use a gold pan, but unless the source vein is intact and reasonable in size, you could be wasting your time.

Metal detectors take a lot of the work out of tracking pocket gold. What used to take days can be accomplished in a few hours with a lot less work. With a detector you can quickly trace the route that gold took as it eroded out of its source and down the hill. Working pocket gold with a detector insures that the pieces you get will be worth the effort. Detectors will not signal on fine or flour placer gold unless it is substantial in quantity and tightly packed together. This packing would be a rare occurrence — one I've never encountered in my years of prospecting. Nor do I know anyone who has found gold like this. A detector will sound on a small stringer of gold in its rock matrix, or a concentrated mass of fines in matrix.

Pocket gold is rough and often attached to mineralized pieces of its original rock matrix. Pocket gold may be stained with various oxides and not recognizable as gold. In this case it must be cleaned with acid later. The good news here is that your detector will recognize stained gold despite the oxides, and unless you get a signal, you do not have to dig. The old "pan" method required the digging of a lot of unnecessary holes in order to find the gold deposition pattern.

The search for pocket gold usually begins along a water course or dry wash. Begin to search for a deposition pattern. When you find gold, continue until you get no more signals. At this point your search will likely begin to take you up a hill. Search in a zigzag pattern, noting where you get signals and digging them to be sure you are still on the gold trail. You eventually find that you are being pulled higher up the hill towards a source and that your pattern of pocket gold is tightening.

I have a shortcut that usually gets me right to the source. Pocket gold will come out of quartz or other rock matrix material. Once I have located my first pieces of pocket gold, I note what the matrix material, if any, looks like. Then I just begin to follow the "float" (matrix material scattered over the ground surface) up the hill. This is easier than following the gold and digging signals as you go. Once I find the source of the float (where it is eroding out of the hillside) I begin a more intensive search and begin digging signals. If the material is good, I may reverse course and search more intensively for buried pocket gold over the face of the hillside. The farther you proceed from the source, the less gold you will find. Pocket gold follows a path determined by gravity, hillside gradient, and water action.

The advantage to my "reverse method" is that once you know the source, you can often, and quite accurately, project, or "read," what route the gold took in its descent as it was pulled by gravity and water flow. Topographic features will also play a part in determining the route the gold took, or where it might have stopped, as it fell away from its source.

Most often the pieces of float with gold attached will be buried, due to the specific gravity of the gold contained within the matrix. Float without gold generally remains on top of the soil surface and travels further down the hillside. There are always exceptions; your detector will infallibly ferret these out for you.

Once you reach the source, a rock pick will be needed to break apart the decomposed vein portion and get at material in situ. Your greatest area of enrichment will be the front part of the decomposing vein and fanning out from it. As you get into the intact area of the vein, enrichment will lessen. Then you begin to get into hardrock mining and part company with recreational mining. If your vein has visible gold you might want to work it for specimen pieces. These sell for much more than the market value of gold alone because of their uniqueness.

Collectors are always looking for nice specimens and you can make some money to buy a second detector — the one that your spouse has been dreaming about since he admired all the gold that you found with yours!

A shovel may be needed to unearth parts of the vein material. Should the vein disappear into solid rock, you have a problem. It will be nearly impossible to follow without heavy rock equipment and dynamite. If the vein looks good, you might want to slap a claim on it

A magnet on the end of your pick will retrieve ferrous (iron) targets that your detector might sound on.

(more on this later). Don't try to fund your lode discovery yourself. It will take all your assets and then some, and you will most likely come out on the losing end. I believe it was Mark Twain who said something like, "A mine is a hole in the ground with a fool standing over it" (actually I think the word he used was "liar" instead of "fool," but sink your money into a lode discovery and you may feel like a fool later!)

In the chapter, "More about Lode Gold" we will get into what you can do if you find a really dandy lode spot. We'll also show you how to clean up those great specimens you find. However, my suggestion is to cover it up so that no one else finds it. Even with a claim, there are those who will "jump" your claim and work it in your absence. If you can't get it all out in one trip, cover it up and disguise the fact that you were there. That way it will be there when you return for more.

Pocket gold is, for the most part, an untapped source of profit for the recreational prospector. Most prospectors just don't know much about it; fewer still bother looking for it. Locating pocket gold with a detector is a cake-walk compared to how the old-timers used to look for it. This brings me to a reminder: as most gold leaves its origination site, it will be found deposited on gradients that drop 30 feet to the mile. Remember this one from the earlier chapter on "reading the river?" It goes for dredging, too!

Recreational Gold Prospecting & the Environment

In this age of environmental concern many outdoor hobbies are under scrutiny. Recreational gold prospecting is no exception. Arguments are ongoing between those whose outdoor interests place them at odds with those who profess deep concerns for the preservation of natural areas. Both of these groups often find themselves squaring off against the government agencies whose job it is to oversee and manage the areas at the center of the land-use controversy.

It is possible to come to some compromise on these issues. Both factions of the debate love and appreciate our beautiful natural resources. The differences seem to be in how that appreciation manifests itself as a way one chooses to enjoy the land.

Where things go awry is that most of the public associates recreational prospecting with large-scale mining. What many environmentalists envision when they hear the words, "prospecting," or "recreational mining," are large gaping holes, tracts of denuded, chemically burned lands, and lifeless, poisoned water. In other words, they envision what they know of mining — which happens to be true of commercial mining before any environmental dangers were considered. This association that has proven detrimental to the recreational prospector.

Hydraulic mining was devastating to the environment.

Recreational gold prospecting is not, in any way, similar to historical commercial mining efforts. We do not use dynamite or cyanide. We do not strip the land of its flora or poison the water. We do not leave large tracts of land barren of vegetation. Nor do we use mercury in our gold pans or gold recovery equipment, as was done during the Gold Rush era, when the effects of mercury poisoning were little known. We also do not move more earth than nature can erase during a single season of heavy rain. Recreational gold prospecting, viewed from a realistic perspective, has no damaging effect, long-term or short-term, upon the environment.

The answer to the dilemma of prospectors and environmentalists is education. Educating environmentalists to what recreational gold prospecting is really all about, and in turn listening to, and learning from them, will serve in bringing peace to our common table.

Part of the problem that prospectors face is some members of our own group. In practically any portion of the population there are members whose actions bring hardship upon the rest through lack of consideration, respect, or failing to accept responsibility for their actions, and the effects of those actions. We have a responsibility to one another, to the environment, to our children, and to other life forms. We also have a responsibility to our hobby and our fellow prospectors.

What all this means is that whatever we do to others and to the en vironment, we are also doing to ourselves and our children. Everything that is done has an impact on the entire planetary community. That's just the way it is. There are too many humans on the earth for us all to unthinkingly do our own thing without regard to its long term effect.

Recreational gold prospecting certainly does not endanger the future health of our planet as much as littering on a global scale, war, air and water pollution by industries, overpopulation, nuclear weapons, overfishing the oceans — the things that our modern cultures take for granted each day. Although these things are creating havoc with our environment, it is simpler to focus attention on a group of hobbyists than to take on the world powers and commercial concerns that are behind the real environmental hazards. And maybe it is easier to meddle with someone's weekend prospecting trip than to alter the life-styles of nations, or of ourselves. Those life-styles are major contributing factors to pollution, degradation of the flora and fauna of the planet, and to the increasing ill-health of our current and future generations. Those things stated, let us examine what environmental impact recreational prospecting actually has, if any.

One of the most controversial aspects of recreational prospecting is dredging. Dredging displaces areas of river bottom wherever it is done. Some see this displacement as an alteration of a natural environment and a threat to fish habitat. While dredging does reorient sand, gravel, and rocks as they pass up and over the sluice bed before tumbling back into the river, does this in fact constitute an ecological hazard? Quite to the contrary, no harm is done to the ecology of the river.

The truth is that a river bottom is naturally reoriented on a much more massive scale by the forces of nature itself. Flooding occurs on a cyclic basis as part of natural weather patterns, and this flooding creates river bottom displacement on a scale far more extensive, and along many more miles of a watercourse, then dredging by recreational prospectors ever could. This fact alone makes the anti-dredging argument not only inconsequential, but ludicrous.

In our chapter on **"Where to Find Gold & Why It's Found There"** we discussed the ways in which gold is deposited and concentrated, and how during certain seasons of heavy runoff, gold is churned up as rivers disgorge their own sediments. Dredging is only mimicking on an infinitesimal scale what nature does seasonally on a far more imposing scale. Yet no one accuses nature of being ecologically unsound! Nor does any one attempt to regulate, legislate, or impose laws to restrict the workings of nature. Humans do this to one another.

Dredging, or any other form of recreational gold prospecting, does not upturn river beds for miles along their length. Nor does it devastate vast tracts of river bank, ripping up trees and shrubs, casting them out upon newly created flood plains to wither and die. Nature does this. Mud slides move tons of earth and the face of the land is changed. Nature recovers and is renewed by these natural occurrences. Plants reseed and establish themselves in new areas. Fish populations continue despite flooding and massive displacement of river bottoms. Animal populations are only minimally impacted by nature's wild ways and seasonal rages. The apparent devastation of forest fires has been found to actually be regenerative to nature, as have the effects of volcanic eruptions. New life abounds in these areas within a short time.

Nature heals itself. And all of this is rightly seen as natural. Few raise the cry of environmental desecration and damage, for these are acts of nature, imposed upon nature by nature. The desire to legislate an eventual elimination of dredging, or to make it difficult to obtain a permit, is based on a lack of understanding of both recreational prospecting and natural earth processes.

Dredging has been restricted to a seasonal activity in order to enable fish to carry out their life-cycle, and this is reasonable. This restriction serves dredgers too, due to the fact that the seasons of May to October are warmer, snow has melted, and water volume is down. Dredging is a much more pleasant occupation in summer months.

The argument over whether dredging damages the environment becomes unreasonable when we consider the fact that with just a minimal rise in water volume, rivers are swept clean of any underwater depressions created by dredging. There are positive impacts of dredging on the environment that are overlooked or ignored.

The river bottom depressions created by dredging serve as homes for fish, which claim and vigorously defend them as their own territories — at least until the depressions are erased by natural seasonal cycles. Fish do not care if their homes are created by human actions or by nature. That concept is an abstraction of the human mind. Fish will just as happily colonize a man-made reef as a natural coral reef — or a depression left by dredging.

Dredging is not damaging to the fish themselves either, but serves to stir up the sediments, freeing food which they fearlessly dart after. Any dredger can attest to the schools of fish swarming around ongoing dredging operations, partaking of the feeding opportunity created.

Dredging, and its impact upon fish populations, was the subject of a study done by the California Department of Fish and Game during the 1980s. The results unequivocally show that dredging is, for the above stated reasons, beneficial to fish populations!

Dredging is often subject to restriction in some areas, based upon dredge capacity. Yet no dredge used in recreational prospecting could have an effect upon the environment that isn't exceeded many times over by natural earth processes occurring on a cyclic basis. Only unrealistic arguments can be used to counter this truth — arguments based in a lack of knowledge of what recreational gold prospecting is all about.

Again, when some in the public sector think of dredging they envision the massive, bucket-line dredges that were operated by commercial mining companies in parts of California and Alaska around the turn of the century, and in some cases until the early 1940s. Some of these massive dredges dug the river sediments down to 50 feet! The effects of this large-scale dredging can be seen when driving through Oroville on Highway 70. Huge piles of stones line the Feather River and can easily be spotted from the road. This was not the work of recreational gold prospectors, but commercial mining concerns that operated decades ago.

What many fail to realize is that nature is in a constant state of flux, ever changing, constantly altering the face of the land and the shapes of continents. Any elementary geology student knows this. Yet there are those who would legislate that not a stone be moved without a fine being imposed. If one is willing to condemn recreational dredging as harmful to the environment, one must also be willing to state that nature itself is bad for the environment!

Since humans tend to see the world as opposites — good or evil, black or white, friend or foe, this or that, us or them — recreational prospecting is associated with historic mining practices which, due to ignorance and lack of technology, were indeed harmful to the environment where they occurred. Today, technology and science have enabled large-scale mining to operate with significantly reduced impact, and enhanced social consciousness has made restoration of environments the end-stage of most commercial mining efforts.

Let us discuss sluicing and highbanking. Here again any impact upon the environment is negligible. The residual diggings of prospectors on the banks and bench areas — the results of this type of prospecting — are totally obliterated by nature with the occurrence of cyclic, heavy runoff periods.

Would anyone who is rational restrict a desert fox from digging a borrow? Part of the problem is that many who want to save the environment make the mistake of seeing human beings as apart from, or alien to, that environment. We are as much a part of nature as the red fox, or the coyote who digs her burrow into an embankment. Yes, there are many things that people can do in the environment that are harmful, but digging is not one of them. Digging is quite natural, whether it's to build a home or to extract gold.

Drywashing depressions left in desert washes are swept away during periods of flash flooding — a natural component of desert weather cycles. The depressions caused by drywashing operations may remain for longer periods of time because of more intermittent periods of flooding in desert areas. Yet, in time, the desert too, sweeps away these minor vestiges of a prospector's presence.

It has been argued that the depressions left by drywashing operations are harmful to desert tortoises, which are considered to be an endangered species in the western deserts. This argument seems logical to the uninformed, until one actually observes how tortoises use these depressions. They make homes in them, despite the argument that desert tortoises fall into drywash depressions and are unable to crawl back out.

On a prospecting trip to the Red Mountain area of California, I spent a week drywashing a rich gold area and was privileged to observe and enjoy the comings and goings of one fearless tortoise. This tortoise had dug her burrow into the side of a four-foot deep hole — the result of a drywashing operation. The sides of the hole were quite steep and she had dug her burrow about midway between the bottom of the hole and its top. She would slide into the hole and, although it would at first appear that the sides might be too steep for her to exit, she had no difficulty in doing so. She would scale the sides of the hole like a little bulldozer.

On one occasion the tortoise tumbled into the hole and landed upside down. Now it has been said that a tortoise upside down cannot right itself, so we observed her predicament to see if this was true. We had every intention of coming to her aid had she been unable to turn right side up. This intrepid critter managed to do just that — with some flailing of her legs — in just a short span of time, and go on about her routine.

In most areas where recreational prospecting takes place, I have seen evidence that other desert creatures seize upon drywash depressions as the raw material from which they too create burrows. Desert foxes, hares, reptiles and serpents all use these man-made depressions and holes as ready-made plots in which to dig a home. Animals tend to gravitate to drywash depressions probably because it saves them some digging in the hard desert soil, and they instinctively recognize the housing potential inherent in both natural and man-made depressions.

Historic mining areas are often depicted to illustrate the detrimental impact of mining, effects which may still be ongoing. Yet there is no connection between these historic efforts and those of modern-day recreational gold prospectors.

I have visited many old mining areas and been enchanted with the wildlife and the variety of plant species to be found there. While these areas are said to be wastelands, the evidence — in most cases — is that nature has reclaimed and is living in harmony with the old workings. Yes, there are some areas still remaining were ground water was poisoned by chemicals used in historic mining operations and wildlife has not returned.

Yet in most historic mining areas I visit, the old workings provide no more than a glimpse of our history and provide a nostalgic atmosphere that speaks of times past. What of the old mining areas of California's Mother Lode, Bodie, Calico, Julian; Nevada's Virginia City, or Colorado's Cripple Creek and Leadville? What would draw people to these places today if it weren't for their mining history? People come to these places, and support them economically, because of the mining history to be seen there.

Water cannon, such as this Hendy Giant, were used to blast apart the sides of gold-bearing hills and mountains.

Our history gives us a sense of who we are. But let us not condemn the present, because of some of the sins of the past. Yes, in some cases historic mining efforts did wreak damage, but let us not condemn recreational gold prospectors because some people make an association between historic mining and recreational gold prospecting. Our target — gold — may be the same, but the methods are not, nor are the effects upon nature.

Legislation has been proposed to would limit metal detecting, another important aspect of recreational gold prospecting. The impact of metal detecting is the most minimal of all modes of prospecting. The target of detecting is, of course, gold nuggets. The holes that prospectors dig to recover nuggets are rarely more than three or four inches in diameter and usually no more than 15 inches deep.

What has caused a problem for metal detecting enthusiasts is the "coin shooting" aspect of the hobby. Coins are found in places which people frequent, such as public areas like parks and other gathering spots. The problem occurs when an enthusiast digs a target out of a lawn area. There is a knack to removing a target without killing the grass and damaging the beauty of the lawn. A few thoughtless enthusiasts have brought scrutiny upon the thousands who engage responsibly in metal detecting hobbies.

In some cases privately owned areas have been damaged by thoughtlessness and disregard for others' property. Although few in number, irresponsible individuals make the rest of us suffer for their negligence. Because of a few careless people, those who use detectors to prospect for gold, recover coins, or hunt for buried treasure, now have to face a future of increasing restriction and legislation of the hobby.

The addition of mercury to gold recovery equipment is not done in recreational gold prospecting. Mercury is a naturally occurring substance that is associated with the mineral cinnabar (HgS or mercury sulfide). When cinnabar is subjected to heat, droplets form and yield the liquid metal we know as mercury. It is the only metal to maintain a liquid state at normal temperatures.

During the gold rush era it was common to use mercury in sluices and gold pans to recover fine gold. Some of this mercury naturally escaped into the rivers and, although rare today, a prospector will occasionally find a piece of mercury-coated gold in his gold pan.

Mercury-coated, or "mercury-gold" as it is called by prospectors, is metallic-silver in appearance. It will behave in the gold pan the same way normal gold does and is a consequence of historic mining processes. Mercury-gold is most likely to be found after a season of heavy runoff when riverbeds have had some of their older deposits upturned and redistributed near the surface.

It was not commonly known by historical prospectors that mercury was poisonous. Many of these old-timers used their gold pans to cook meals, or they used their cast iron cooking skillet to pan gold. Cooking 'taters and biscuits in the same pan in which they used mercury proved to be an unhealthy practice that a number of prospectors engaged in!

When we envision these old sourdoughs we often picture a bony, grizzled fellow with sallow, wrinkled skin, a wild look in his eyes, and few teeth left in his jawbone. The effects of mercury poisoning are many, but some of the more obvious ones are premature aging of the skin, early graying of the hair, and loss of teeth. It is very likely that our mental image of the sourdough is partly because of their ill-informed attitudes toward mercury. Many of these old-timers were in fact suffering from mercury poisoning through accidental or intentional ingestion. As an example, take the "miner's cocktail," a drink composed of a shot of whiskey and a blob of mercury which was usually downed in a single gulp!

Enjoyment and care of nature are integral parts of recreational gold prospecting.

While mercury may still be used to recover fine gold, it is never used in recovery equipment or in gold pans as it was in former days. Today mercury is not allowed to escape into the natural environment. When it is used, it is recovered from the process and reused again and again. There are a few recreational prospectors who do use mercury to process flour gold. This is done at home with care and special equipment. Most recreational prospectors don't bother using mercury.

For those who do there is equipment on the market which allows the prospector to take advantage of the marvelous properties of mercury without subjecting self or environment to its damaging effects. One way that prospectors use mercury is with a retort. A retort sells for around $50. There are other methods for safely using mercury in processing fine gold from black sand, but that is a complex subject of study that should be left to the individual prospector. In the last chapter I have listed some sources where you can find information on the subject of amalgamation — the use of mercury to recover fine gold.

We have covered recreational gold prospecting and the environment and discovered that it has a relatively small impact on the environment. Nonetheless, recreational gold prospectors have a responsibility to our hobby, to our fellow hobbyists, to other humans, and to nature. There are a few simple rules that anyone going into the wilderness, for any reason at all, should observe. It just makes sense.

1. If you pack it in, pack it out. You've heard this one before. Do it!

2. Never dump oil or gasoline into the natural environment, whether in the wild or at home.

3. It is, harmful, risky, and unnecessary to use rocks and trees for target practice.

4. Adhere to campfire regulations at all times.

5. Respect all personal property, "No Trespassing" signs, and claim notices.

6. Respect and enjoy the area you are in.

7. Always take a First Aid kit.

8. Obey the regulations of the governing land agency.

9. Respect the rights of others to be there, too.

10. Don't do anything you wouldn't want done to you. It's all about Respect — self-respect, respect for others, respect for nature.

Planning a Prospecting Adventure

Most prospecting trips will entail some form of camping. It is possible, on some occasions to stay in motels and drive daily to the prospecting site. But this not only robs you of time that can be spent prospecting, it subtracts from the experience of enjoying nature. The enjoyment of prospecting is intimately tied to the experience of being out in nature.

Many prospectors are also RV campers. RV campers have many more amenities than the tent or car camper, yet still enjoy close contact with nature. Gold prospecting and RVing combine well together — until you find a prospecting spot that can't be accessed by the RV. At this point, you may find yourself shopping for a small four-wheel-drive vehicle to tow behind the RV. However you choose to camp, there are many ways to make it an utterly delightful, satisfying, aspect of your gold prospecting.

One chapter is not sufficient to cover the subject of camping in its entirety, but I can pass on some tips to make camping a more successful and enjoyable adventure, and a pleasant adjunct to your prospecting trips. There are many books available on camping for those who need more information.

A successful prospecting adventure requires careful preparation and implementation. There is nothing more frustrating than to get out into the field and discover that you've left an important item at home. A successful prospecting trip is a combination of research and organization. That's why I leave nothing to chance.

First and most important is research. If you haven't researched whether a spot has gold potential, you could waste days or weeks of your valuable time. There remain numerous spots where gold can be found, but there are also countless spots where gold is not found. Through research, you can place yourself in the best possible locale for finding gold. Sampling techniques then narrow down your search to the richest concentrations of gold in known gold areas.

The best time to research alpine gold sites is when the snows lie heavy in the mountains and the only prospecting that can be done is "armchair" prospecting. Winter is the time to visit libraries, send for maps, read, research, and plot the coming spring and summer prospecting trips.

After snow has melted and the water volume in rivers has dropped at higher elevations is the time to head out for some "early bird" prospecting. Those who plan to crevice, nugget shoot, or sluice for gold, can get a month or month-and-a-half jump on dredgers, who must await the opening of dredging season around the beginning of May.

We in the western desert of the U.S. do most of our research during the summer, when searing temperatures provide us with the opportunity for armchair research. Summer soon gives way to late fall and winter when desert temperatures begin to drop and glorious, mild sunny days are perfect for drywashing and nugget shooting gold-rich desert washes and alluvial placers, or detecting around abandoned miners' cabins for relics and forgotten pokes full of gold.

Once I've done my research, I spend a day or two in the garage going through my camping and prospecting equipment. Once I know where everything is and its condition (does the camp lantern need new mantles?), then I decide what vehicle (RV or 4x4) the proposed prospecting site will accommodate.

Many of my favorite sites require a four-wheel-drive vehicle, so it may be time to check the tent for any needed repairs that may have gone unnoticed from the previous season. Many of us tend to stow the camping equipment carelessly in the garage after the last trip of the season. I do this on occasion, then have to spend a day or two organizing what I neglected the previous season. Often, especially if I'm going prospecting alone, I leave the tent behind and sleep in the car.

If my proposed prospecting site is in a public area, perhaps I can take the RV instead. This of course requires a different kind of preparation. Each type of camping requires its own packing and planning. Whichever vehicle I plan to use, it is cleaned, the oil is checked, and the needed prospecting and camping equipment is prepared.

Two weeks before the trip I begin to make a list of all the items I plan to take. After a season of disuse, you will likely not remember just what items are in the camping bins! There is usually something I forget to put on the list, and that is where the Prospector's Notebook comes in.

My Prospector's Notebook is a small ring-binder in which I keep a multitude of useful information. I have sections for lists of "must take" items for tent camping, RV camping, or 4x4 camping, and lists of the gear needed for my various prospecting methods. Also in the notebook is mileage and directions to my favorite (and sometimes secret) campsites. Range and township numbers on topo maps that pertain to prospecting sites are also included, along with all types of field notes. Mileage is important to have recorded in your notebook, especially when retracing your steps in the desert.

Too many times when I've been to a great prospecting site and can remember generally how I got there, I forget the precise spot. Prospectors (and others) are famous for forgetting these all-important facts — hence all the lost mine tales! My Prospector's Notebook has helped me return to great spots, even after long intervals of time have passed since my last visit. I also collect other minerals in addition to gold, and like to keep track of the mineral and fossil locations that I happen upon when prospecting.

When I camp and prospect I use the "bin" method of organization. At home-improvement stores you can purchase plastic bins in a variety of sizes and colors. I use them to organize my camping and prospecting gear. In one bin I put kitchen and camp utensils. Here is a list, from my prospector's notebook, of what goes in this bin.

1. Tea kettle.
2. Frying pan.
3. Pot with lid.
4. Paring knife, chopping knife.

5. Spatula & wire whisk.

6. French whisk.

7. Wooden spoon & can opener.

8. Fork, spoon, knife.

9. Plate, bowl, cup.

10. Cutting board.

11. Aluminum foil.

12. Paper towels.

13. Dish soap (doubles as shampoo & body wash soap).

14. Pot scrubber, sponge.

15. Olive oil, salt & pepper, sugar, chili flakes, cinnamon.

16. Tea, hot chocolate, coffee.

17. Powdered milk.

18. Hot pads.

19. Plastic wine glass and cork puller.

20. Matches.

21. Toaster rack.

22. Charcoal & small barbecue.

A general camping bin would contain:

1. Camp lantern & mantles.

2. Sewing kit.

3. First Aid kit.

4. Duct tape (endless in its applications).

5. Tow rope.

6. Ax.

7. Heavy cord.

8. Flashlights.

9. Saw.

10. Whisk broom.

11. Collapsible bucket.

12. Candles.

13. Rain gear, warm gloves, knit caps.

14. Repair kits (tent, auto tires, etc.).

15. Steel wool.

16. Dish washing tub (can also double for panning).

A bin for prospecting equipment would contain:

1. Gold pans.

2. Hoe pick, rock pick, chisels, gad bars, sledge hammer.

3. Crevicing equipment.

4. Vials and canisters for gold.

5. Gloves.

6. Screens.

7. Extra detector coils.

8. Short shovels & trowels.

9. Spare batteries (for detectors).

10. Belts, oil, spare parts for recovery equipment.

11. Tool kit.

If you use bins to organize your camp equipment you will always know where to go for a certain item, and so will everyone else with you. Bins come in a variety of colors, so you can color-code them according to what they contain. If you are car or tent camping, the bins can be removed from the vehicle and stowed underneath it, or covered with tarps to protect them from dew and rain. Decamping is easy with bins as long as everyone is conditioned to return items to the proper bins when finished using them. The next morning, just stow the bins in the vehicle and go.

Tarpaulins are endlessly useful when camping. They can be rigged into a tent, a lean-to, or sun shade. They can be used as a ground cover beneath a tent or secured over a tent as a rain fly. One use of tarps is for getting a vehicle unstuck from sand. They work far better than strips of carpet or screen, or packing the tires with vegetation. They are also easier on the environment then the foliage option (which should only be used as a last resort).

In planning prospecting trips, we all have some varying requirements. Many people say that they hate camping and probably recall some grim trips to rationalize their dislike. Camping can indeed be a merciless activity if you go without considering the following:

1. Personal and auto safety.

2. Drinking water.

3. Comfort, warmth, and meals.

4. Special needs of kids and pets (include your own needs here too!)

5. Camping location and conditions.

6. Where to restock groceries and fuel.

Your automobile is your lifeline. Before a trip check the hoses and belts. Pack along extra hoses and belts in case you need to change them. A spare tire, in good condition, is a must, as are the tools needed to change it. Tire resealing kits are not a permanent fix, but can get you to a repair station. Tools to make simple repairs are a must. Duct tape is useful for a variety of temporary quick fixes. Bring extra water in case a radiator hose breaks. For desert trips it is recommended, when possible, to take two vehicles into remote areas. Always notify someone at home where you are heading, how long you will be gone, and whom, or what agency, to call if you don't return at the stated time. For desert prospecting, extra water is a must. It is advisable to dress in layers so that as temperatures increase and decrease you can dress up or down as required. Knit caps and gloves will help greatly to conserve body heat.

A cellular phone can be of great help on remote prospecting trips, particularly desert trips. Most of the places I prospect in the desert can be reached with a cell phone. However, if a mountain range lies between you and a major highway the cell phone may be useless until you get around, onto, or over that obstacle. Alpine areas are sketchier when it comes to cell phone service. It is possible to purchase satellite telephones for around $8,000 and never be out of touch no matter where you are. But just wait. Before long, cell phones will be operating from satellite link-ups and there will be no limit to where you can dial out from. Then cell phones will become major assets to adventurers of all types, at a fraction of the present cost.

Another piece of equipment that you may want to invest in for is a Global Positioning System, or GPS. This piece of equipment uses earth-orbiting satellites to closely pinpoint your current position, which is indicated to you via the readout. The price of these instruments has fallen since their introduction several years ago. A GPS may be purchased from around $300 up to $1,000, or more. The GPS calculates your position based upon readings taken from 8 to 12 satellites, depending on manufacturer's specifications. A GPS is accurate to approximately 100 yards, due to current military scrambling of the signals, supposedly for security reasons. Within a year or so, this scrambling will be eliminated and you will get a true reading on your exact position. A GPS would make relocating that featureless, but rich desert placer, a sure bet!

A well-planned and executed camping trip does not need to be cumbersome to be effective. I recently spent 18 months in the field —

in approximately six-week intervals — to research a book I wrote on gem, rock, and mineral collecting entitled, *Rockhound's Guide To California*. I did the research on my own, packing everything — and I mean everything — that I needed for a series of six-week stays in the wilderness, into my Suzuki Samurai 4x4. I removed the back seat and stowed one or two five-gallon water containers, five gallons of gasoline, rock picks, collecting bags, sleeping bag, pillow, sluice, pans, detector (I did some gold prospecting on my research trips), cameras, notebooks, maps, sand chair, one camping bin (with the barest of essentials, ice chest, auto repair tools, etc. I didn't pack along a tent because the passenger seat reclines, allowing me to sleep in the car. Too, I was seldom in an area for more than a day or two, and a tent would have been a lot of trouble to set up and dismantle every few days. Along with all this I packed canvas bags for mineral specimens I collected. This sounds like a lot to pack into the back of a small vehicle, but organization will allow you to take much more then if you cram everything into your vehicle "slap-dash."

I also packed mineral guides, food, and drink. Although I forwent many luxuries, I allowed myself some comforts to make the trips bearable. Good coffee, fresh food, and a glass of wine at sunset are personal pleasures that I enjoy at home as well as in the field. Even on an austere trip, done for purposes other than pure pleasure, these simple items can make all the difference.

Examine your home life. What are the small luxuries that you savor? A crossword puzzle in the morning? A cup of tea or hot chocolate in the evening? A pipe after dinner? A morning cup of espresso? Whatever your small pleasures, incorporate them into your camping/prospecting trips. You will find that small luxuries and pleasures will offset the lack of the major luxuries — the bathtub or shower, the Jacuzzi, the feather bed, a roof over your head. Sleeping on a cot will not seem so rugged if you can arise to the aroma of freshly brewed cappuccino!

It is possible to purchase tiny espresso and cappuccino makers designed for camping, but I have found that some of the instant cappuccino mixes are excellent, rivaling very closely what I can concoct at home. Also, those individual servings of coffee that look like tea bags are great for camping; the coffee in them brews up to be quite tasty.

If you must prepare a cappuccino, here is a great tip. Pack along some powdered Carnation nonfat milk. It mixes up with a minimum of fuss and mess. When mixed with water and frothed by spinning a wire whisk in it, you will get the dandiest, fluffiest "steamed" milk to adorn your coffee. You can also pre-grind those gourmet coffee beans that you adore and brew them in the field.

The old timers used to brew their coffee in a cast iron skillet. Time and practice will allow you to master this method and turn out a good, strong cup of coffee. You can always toss in a small coffee pot for boiling water for tea and cocoa, and for brewing coffee.

I usually tote along a bottle of burgundy, or other wine that is best served at room temperature. This way I don't waste valuable ice chest space cooling a bottle of wine. While you may be sleeping on a cot, or a pad on the ground, sitting in folding chairs, and spot bathing, you will enjoy your camping so much more if you can incorporate such small luxuries into your camp routine. Even those who claim to hate camping will find that it can be a pleasurable means to an enjoyable and profitable end — that of recreational gold prospecting.

No matter where you camp, you will need to take some water. Desert camping requires even more water for drinking, bathing, for the automobile (if need arises), and for panning. The desert prospector is wise to recycle and conserve as much water as possible. Recycle water used in personal bathing, or in kitchen washing, into the panning bin. You can also dump tea or coffee beverages that were not consumed into the panning bin. Many alpine areas have water available that is at least useful for washing and panning, but you may still wish to bring safe drinking and cooking water.

One handy way of carrying along extra drinking water is to have it serve double duty. I purchase bottled water, or refill the bottles and freeze them at home. These bottles of ice are then used in the ice chest instead of block ice. When the ice melts it is drinkable water, unsullied by contact with food in the ice chest.

Water can be saved by using paper plates and cups. These can be used, in turn, to start the evening campfire. Burning plastic plates and cups releases toxic fumes into the air. For starting an evening campfire when kindling is damp or scarce, I always keep a pack of steel wool in the camp bin. Speaking of campfires, always have a permit on hand for yours, especially if you are camping on land under the jurisdiction of the U.S. Forest Service. Permits are inexpensive and can be obtained from any Forest Service office. A permit is not usually necessary if you are in a campground where fire rings are provided. Always check whether or not a fire permit is required where you will be camping.

When camping I tote along a small three-gallon ice chest. Into it go perishable items such as cheese, meats, and dairy products. In desert environments my bottled ice lasts from two to five days, depending on the temperature and time of year. I use my perishable food items first. Once the ice in the bottles is melted I have a new source of drinking and cooking water. Then I depend on less perishable items for meals. I seldom take along canned foods or those little packets of freeze-dried items sold for backpackers. What I do take along is a mesh bag containing some onions and garlic, oranges, apples, and potatoes. I take along some previously prepared cornbread or muffins. I may tote along some boxed rice or pasta dishes, as long as the only ingredient I need to prepare them is water.

Potatoes are a great item to take camping. They can be fixed many ways, as can rice or pasta. Scrub the spuds and dry them before leaving on your trip in order to conserve the water that would be needed to wash them in the field. Fried potatoes and onions are a dandy treat after a day of prospecting — real comfort food. Pasta or rice can be cooked and tossed with a bit of olive oil, salt, chili flakes, and minced garlic. Apples and oranges make great fresh snacks. Apples can be cored, seasoned with cinnamon and sugar, wrapped in foil and placed near the hot coals to bake for a sweet, satisfying desert that smells like freshly baked apple pie.

I try, when prospecting and camping, not to bring anything that is not absolutely necessary. Therefore, my prospecting buckets do double duty. Placed upside down next to my camp chair, they make perfect little side tables. They also make nice, if not totally comfortable, stools if unexpected guests drop in.

I enjoy lounging around camp after dinner and watching the sun set over distant hills, and the emergence of evening creatures as they begin to stir and search for food. I don't usually pack a radio, TV, or reading material when I prospect/camp as I have found that nature's panorama and sounds are what I enjoy. I can enjoy electronic entertainment and reading when at home. I do take along a small, hand-held weather radio. This has proven invaluable as a warning for incoming inclement weather. If rain is predicted I will make my camp on high ground. For wind advisories, I will seek low spots.

When in the wilderness I like to immerse myself in the experience of nature. It is very healing and stress-relieving to do so. The sounds of wind and water, the rustlings, peeps, and songs of night creatures have the power to take you away from the stressful routines and demands of everyday life; I like to give myself over to this experience. Also, I have found that my sunset musings have often led to inspiration on where to prospect. When the mind is at rest and allowed to harmonize with its surroundings, insights and inspirations may reveal themselves that would not otherwise in our normal, preoccupied modes of thinking. In a relaxed state of mind I may notice an aspect of a wash — unnoticed previously — that looks like a good bet for finding gold.

Camping goes hand-in-hand with recreational gold prospecting. Both of these wonderful activities take us away from the clock, the telephone, and everyday responsibilities. Camping and prospecting are priceless gifts that you give to yourself and can share with your family and friends.

A love of nature is a precious heritage that you can instill in your children — a legacy that will be passed from generation to generation. Enjoy!

Gold Prospecting with Kids

Gold prospecting with the kids is a great family activity. Our lives are fragmented by the various demands that are placed upon us— the need to work, our church and social activities, the need of kids to go to school or be involved in sports. While all of these things contribute to the growth of personality, knowledge, and moral responsibility, they do reduce the amount of time that families spend together. Too, TV programming tends to segregate the time families spend together, especially when so many kids have their own television sets. All of these things separate family members, rather than bringing them together. With so many obligations, adults either may not have the time to play with their kids, or may have forgotten how to.

Recreational gold prospecting, and the camping that it entails, is a wonderful, fun way for family members to bond and play together. On prospecting trips family members are "partners" — in having fun, relaxing, and sharing the adventure of gold prospecting. It is an honored tradition that everyone on a family prospecting trip shares the gold that is found. When my nephew, Perry, reached the age of six, I began to take him on prospecting day-trips to the gold-rich San Gabriel River in southern California. I taught him how to pan the concentrates of the sluice box and how to spot gold in the sluice's nugget trap. As he got older he began to help with digging, carrying buckets, and dumping gravel into the sluice, as well as panning concentrates. With great satisfaction he would collect the gold from his gold pan and secure it in his vial.

Ready for a day of nugget shooting.

Gold from these trips became a valued item for school "show-and-tell." It also became a highly regarded item of barter with the kids in his neighborhood. When Perry grew older I took him to northern California, where we visited historic Rich Bar, under private claim to Norm and Mike Grant. Norm and Mike open their claim to the public on a "fee-panning" basis. Perry was excited to recover some nice nuggets from the gold-rich gravel terraces at Rich Bar. Now that my niece, Allee, is expressing an interest in gold prospecting, both Allee and Perry join me on day-long prospecting trips. We also enjoy the occasional treasure-hunting trip to a local beach.

Prospecting with these two youngsters has been a rewarding bonding experience for all of us. Kids have an exuberant response to nature. Watching them imbibe our shared experiences makes me feel young and reminds me of what it is to play again. There is much to excite and tantalize their innate curiosity and, through their exuberance, adults can experience the world in ways we have forgotten. Gold prospecting provides both physical activity and mental stimulation for kids and reminds adults to relax and loosen up. For kids, the challenge and adventure of finding their own gold and showing it to friends is satisfying and thrilling. Gold prospecting and camping are activities that take them away from their usual routine of school, sports, TV and video games, and introduce them to nature.

When prospecting with kids, there are some things to keep in mind. First and foremost is safety. The younger the children, the more they will need to be watched. Prospecting is done in areas where natural dangers abound. Rivers, especially, are hazardous for young children, who must be constantly supervised by a responsible adult. Deserts are rife with other hazards such as rattlesnakes and abandoned mine shafts. Young unsupervised children could be at risk from coyotes or the occasional cougar.

Prospecting trips with young children must be geared to their activity level. These trips will be more for fun and benefit of the kids than for the recovery of profitable amounts of gold. For really serious prospecting it is best to leave young children at home or in the care of someone else, while the adults indulge in some really earnest prospecting.

Next comes their comfort. Kids need to be kept warm, dry, and comfortable. If they have a favorite toy, game, or snack, allow them to bring it. A child, particularly a younger child, will take much comfort in having a familiar toy to cuddle up with. Next, patience and attention to their needs are the greatest assurances a parent can give a young child. A youngster who has never been camping will need reassurance. Strange sounds and sights may at first make youngsters uncomfortable. They will take many nonverbal cues from the adults and older kids in the family group. If the adults are relaxed and having a good time, the youngsters will feel confident and secure. A combination camping and prospecting trip with youngsters' needs and activity levels foremost in mind is a great way to start them on a lifetime love of nature and recreational gold prospecting.

When kids get older they will not require as much supervision and will be able and willing to participate in sampling, digging, and panning. Kids have an excellent instinct for finding gold once they understand the basics. Many of them have a natural talent for panning. Pick a gold pan for a child just as for an adult. Refer back to the chapter, "**Basic Techniques of Gold Prospecting**" for tips on picking the right size gold pan for a youngster. Plastic gold pans may be the best choice for a child, since they require less care than metal pans.

Children spontaneously take to a variety of outdoor activities; some will enjoy fishing while others pan for gold in the same stream.

Panning is usually the first activity that a youngster can learn. Kids develop the coordination for panning around the age of five or six, although I know of children younger who are adept with a gold pan. Young children may be awkward when first learning to pan. This can be exasperating, but patience and praise will go further than irritation and criticism. Remember your own first attempts at panning! Each child will be unique in his or her aptitude for prospecting. However, if the children are not interested in gold prospecting, they should not be pushed. A child can easily learn to dislike something if forced into it. Don't be surprised when children suddenly cease their prospecting activities and engage in some serious exploration of the surrounding area. Just as suddenly, they will return and resume their former activity. Young preteens and teenagers also are inclined to this sudden change of pursuit.

Young children need to be constantly supervised and encouraged to pursue activities with an adult's range of vision. Older kids just like to experience everything around them and will make occasional forays out of sight to see what might lurk just beyond their eyesight. When they return their eyes will be shining with excitement and they will likely tell you of their discoveries and show bits and pieces of rocks, relics, or foliage.

At an early age most kids are enthralled with nature and eagerly bring home rocks, leaves, and insects. This often persists into later years. Recreational gold prospecting fits right into this inherent love of nature which seems to be a universal part of childhood. Kids thrive on activities in which adults appear to be interested in participating. Both child and adult are greatly benefited by this sharing of recreational activities.

Jamie and her mom enjoy panning while dad shoots for nuggets and Danny shovels gold-bearing gravel into the sluice.

Part of the enjoyment that kids get out of prospecting is in knowing something of its history. When prospecting with my niece and nephew, I don't make the history lesson too involved, but do tell them simple facts that they can relate to. For example, I may tell them that the old-timers not only used their gold pans to pan for gold, but also used them as frying pans for cooking meals and as plates from which to eat. My niece and nephew enjoy emulating, with their own gold pans, the old-time prospectors by using their pans as plates for their sandwiches. We rinse any dirt from our pans and sit cross-legged on the ground or perch upon a rock with our lunches in our pans. They also enjoy using their gold pans to dip up a drink of water from the river — if we are in a spot where the water is clean and pure enough for drinking. They seem to greatly enjoy the fact that I too use my gold pan as a platter and a large drinking cup.

These impromptu history lessons on Gold Rush history also provide subject matter from which to draw for school reports. Photographs of the family prospecting enhance any written report a child creates for a school assignment.

The evening campfire is a wonderful way for the family to relax after a day of activity. This is a good time to pass around the gold and get each one's opinion on what to do with it. This is also a good time to divide it up so that each family member has a share of the day's take to admire. If the adults keep all the gold for the purpose of "keeping it safe," this is not nearly as much fun or as rewarding to the kids. And if the kids lose their gold? Much of the fun in having gold is in the finding and getting of it. What is a little lost gold compared to the joy of bonding with your children?

Hot cocoa, and marshmallows toasted on a straightened coat hanger, are traditional pleasures to be enjoyed around the campfire. So is telling ghost stories, but when camping with kids, other types of bedtime stories might be less likely to result in nightmares. Getting away from the city and its nighttime illumination is a good opportunity for the family to become familiar with the stars which make up the various constellations, the planets, and the moon. Children readily take to leisurely lessons on the flora and fauna that abound in wild areas. They enjoy sitting still — temporarily — if it will result in a glimpse of an elusive wild creature.

I have noticed that children's sports are no longer the simple, unsupervised, carefree play that my childhood friends and I enjoyed in vacant lots. Today, children's sports are organized, regulated, and supervised by adults. Children's sports require expensive uniforms and are rife with competition and emotional pressure. But I have found that kids today really do enjoy simple pleasures, despite the fact that their toys and games are far more complex than those of my own childhood. Perhaps children too, feel the need to escape from the ever-increasing complexities of everyday life and enjoy the simple pleasures of sun, water, nature, play, family interaction — and gold prospecting!

What To Do with Your Gold

Once caught by the golden lure you become a prospector for life, condemned, doomed, exalted!
  —Edward Abbey, *Desert Solitaire*

Gold! Its beauty appeals immediately to our emotions and aesthetics; its durability inspires a sense of the eternal. An encounter with gold transfixes us, whether through a jeweler's window, at river's as we espy nuggets lying in the nugget trap of the sluice, or when recovering a nice piece of lode gold targeted by the detector

Gold's rarity makes us want to seek it out all the more, even to hoard it. And gold certainly creates a desire to surround and adorn ourselves with it. The enchantment that gold weaves upon the human imagination is enduring and inescapable.

The gold prospector and miner who seeks and finds gold knows well the allure this precious and beautiful substance exerts on the human psyche. As recreational gold prospectors, we spend a great deal of time in our quest for gold, and often return from prospecting trips with gold nuggets and flakes. However, finding gold is not the end of the story. Once gold has been found, we must decide what to do with it.

I will share what I do with my gold and answer some frequently-asked questions such as, "What do I do with the gold I have worked so hard find?" "Where can I sell my gold?" "What procedures and techniques are used to create gifts of gold?" "Can I even bear to part with my gold — something that I have worked so hard to get in the first place?"

There are many things that you can do with your gold. The first and simplest is to just keep it. Many prospectors keep their gold to admire and show to others. Some keep it around for a "rainy day" when it could be sold for needed funds. Keep in mind, however, that once gold is sold or made into jewelry, from which a profit is obtained, those profits are subject to taxation. Native gold, although valuable, is not subject to taxation unless sold.

You should start some sort of organized collection. Many prospectors choose to sort nuggets according to size, or by location where found, and keep them in water-filled vials. Flakes are also gathered into vials. The gold-filled vials are usually kept in some type of felt-lined storage box. Keep your gold in a safe spot in your home.

Prospectors generally like to remove all traces of black sand from the vials in which they store and display their gold. This is a simple matter when dealing with nuggets, but when retrieving flakes from the gold pan, the small brush that is used often picks up black sand and silica, which end up deposited into the vial along with the gold.

There are several ways to eliminate all black sand from your vials. You can pour the contents of the vial into a gold pan and very carefully separate black sand from gold using your brush, then redeposit the gold flakes into a clean water-filled vial. You can use a magnet to draw black sand away from gold after pouring the vial contents into a **plastic** gold pan. A magnet can also be drawn along a vial held horizontally to draw black sand to the opening where it can be poured out. Just be careful not to pour out any gold!

Since most black sand is made of iron, a magnet will work well to segregate it from gold. However, there will usually be some fine grains of silica upon which the magnet exerts no effect. Delicate panning will completely isolate gold for display from either silica or black sand.

Many prospectors like to "brighten" their gold by putting it into a vial containing white vinegar. The gold and vinegar are left to "steep" together for a couple of weeks. The vinegar may be flushed by running a slow stream of water into the vial. Silica may also be flushed from a vial using this same method, due to its low specific gravity relative to gold.

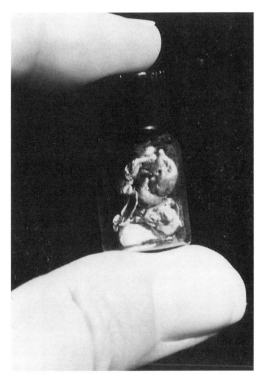

Your gold nuggets are jewelry waiting to happen.

When "flushing," always place a gold pan beneath your vial in case you drop it or you eject too mighty a stream of water into the vial and some gold flakes are washed out. One advantage to storing gold in water-filled vials is that the water exerts a magnifying effect on your gold, making it look even bigger and more impressive when you go to show it off!

A rotary tumbler used for polishing rocks can also clean up nuggets found in dry desert washes. Nuggets from dry placers sometimes need a bit of brightening or oxide removal. Use about one tablespoon of TSP or plain soap powder, add your nuggets, and fill the tumbler barrel with water. Seal it and let the tumbler run about 30 minutes. Rarely do nuggets that are subject to water movement require brightening.

Another simple way to show off your gold — one that does not require any metal or stone-working skills — is to purchase small, clear glass baubles. There are a number of companies that make pendant-style vials and baubles that can be taken apart, lined with velvet, filled with gold flakes or a prize nugget, and resealed. They come in a variety of shapes and sizes and can be used for making earrings or pendants. Pendants, earrings, bracelets, and key chains can be fashioned from the baubles just by adding a length of chain, "shepherd hook" earwires or a key chain. Many prospectors purchase baubles, cut and glue in a black, red, blue or green velvet backing, and add a nugget. Place the bauble on a chain and you have instant and fashionable gold-flake or nugget jewelry. These can be given as valued gifts. Some prospectors buy baubles in quantities and a selection of styles, fill each of them with a small nugget or a collection of flakes, and sell them at swap meets.

Some sources for glass jewelry vials and baubles:

The Craft Market
401 5th Avenue
Fairbanks, Alaska 99701
(907) 452-5495
Also markets a book, *Gold Nugget Jewelry Making*.

Miner's Keepers
P.O. Box 175
Sumpter, OR 97877
(541) 894-2208

Griegers Inc.
P.O. Box 93070
Pasadena, CA 91109
1-800-423-4181
Also soldering equipment, gold wire, jump rings, spring rings, "findings," jewelry making supplies, lapidary equipment.

Making jewelry from your gold increases the value of your finds and assures that you will maximize profits from your recreational gold prospecting. By making and selling your own jewelry it is possible to increase your profits to as much as $1,000 per ounce, or more. This is one of the best ways to increase the value of your flake gold, which otherwise would only be worth its price in Troy ounces.

Perhaps you would like to sell your gold or have it refined into bullion bars. Pure, unalloyed 24-karat gold sells for approximately $383 per ounce at this writing, but the market price fluctuates daily. Current gold prices may be found in the investment section of most newspapers, or on most network news broadcasts.

Gold in the form of flakes will be alloyed with various metals, usually copper and silver, which will reduce its value from the market price of gold. You may take your fines and flake gold to a company that provides precious metals refining services if you must have it refined and poured into small bullion bars. Bullion bars make a valuable addition to any retirement savings plan. Many assay labs will refine small amounts of gold for a fee. Look under "Metals Refining" or "Assaying" in the phone book to locate a service near you. Be sure to contact any company you are considering and determine that they are legitimate before you leave your precious, hard-earned gold with them. A great way to find assayers and metal refiners is through the *International California Mining Journal*. This publication may be found at most newsstands or can be ordered by writing to ICMJ, P.O. Box 2260, Aptos, CA 95001. Also check the classified sections of prospecting-related magazines for assayers and refiners.

Whether gold is found as placer or lode, it is an excellent medium in which the prospector may create, or have created by someone else, beautiful and memorable jewelry, art, or visual displays. Natural gold nuggets — in addition to fine lode specimens displaying visible stringers, wires, or crystals of gold — sell for more than their Troy weight and relative purity would indicate. They may also sell for **more** then the current value of refined gold; choice nuggets and lode pieces have what is termed "specimen value." Nuggets have specimen value because each one is as individual as a fingerprint. Each is molded and designed by the journey it takes from its original source to the place where it is eventually found. Lode gold tells the story of deep, ancient earth processes which may result in a continent or in something as lovely and valuable as gold, trapped yet still visible within its matrix of stone. This creation depends on massive, internal earth-forces of heat and pressure spanning vast periods of geologic time that dwarf, in comparison, the limited human experience. Perhaps this is why collectors pay great sums for excellent specimens of lode gold — to hold, if only for a brief moment, the seemingly infinite.

Metalsmithing and other jewelry-making skills can further increase the value of your gold nuggets. Most prospectors do not have their gold refined before selling it. They usually sell nuggets "as is," or make jewelry from the fines to sell at swap meets or to give as gifts. Fine specimens of lode gold may be sold with minimal cleaning or be carved into lovely one-of-a-kind sculptures. A major mineral specimen dealer to collectors is Kristalle, owned by Wayne & Dona Leicht. They purchase and sell specimens of fine lode gold.

Kristalle
875 N. Pacific Coast Highway
Laguna Beach, CA 92651
(714) 494-5155; FAX (714) 494-0402

Lode gold may also be slabbed, or sawed into slices, shaped and polished with lapidary equipment into cabochons (polished, but not faceted, convex-shaped gems) for subsequent setting into jewelry mountings.

Gold nuggets, found via a recreational prospector's luck and skill, can be transmuted through metalsmithing, transforming native slugs of gold into one-of-a-kind jewelry. The prospector who searches for placer gold will eventually be rewarded by finding material from which to create, or have created by someone else, an item of enduring beauty and value. Many jewelry stores employ their own goldsmiths, and will purchase gold nuggets for making jewelry.

If you are a goldsmith yourself, you may choose to design and fashion your own gold nugget jewelry. Simply by brazing or soldering your gold nuggets to jump rings and chain, earring "clutch-backs" or wire ear hooks, you can fashion unique, personalized jewelry to wear or give as gifts. I have a pair of earrings that I made from a couple of nuggets, related in size and shape, that I recovered from my sluice box while prospecting the North Fork of the Yuba River in 1986. Think about designing your own hat bands, pendants, bracelets, or earrings. As your skill in soldering grows you may want to try creating rings, belt buckles, watchbands, or bolas. Turning your nuggets into lovely, unique jewelry is not as difficult as you might think.

To make gold nugget jewelry you will need some equipment:
1. Miniature (acetylene or propane) torch such as the Smith "Little Torch."
2. Torch tips (sizes "A" or "3" for most jobs).
3. Safety goggles.
4. Flux & brush.
5. 14-karat gold or "hard" sheet solder (1100° to 1600° Fahrenheit).
6. Pickling solution, such as Sparex.
7. Soldering block or asbestos pad.
8. "Third hand" for holding your project while working.
9. Sharp-tip self-locking tweezers, two pairs.
10. Chain-nose pliers, two pairs.
11. Metal snips or lightweight sharp shears.
12. #2-cut, round and flat metal files.
13. Gold nuggets.
14. 14-karat jump rings.
15. Water soluble red rouge.

Begin by making a simple gold nugget chain to wear or use as a hat band. Use your shears or snips to cut the solder into "snippets" about 1 or 2 millimeters in size, depending on the size of joint to be soldered. The size of the snippet will vary with the size of jump ring you use. It is easiest to do this by cutting only a $\frac{1}{4}$- to $\frac{1}{2}$-inch portion of the sheet in strips lengthwise, then cutting crosswise to create tiny square snippets. Set these aside.

Select the nuggets you wish to use by shape, thickness, etc. Like sizes and shapes create a balanced look, but the point is to be creative; design your piece according to personal taste, keeping in mind simplicity, beauty and balance. You will need two jump rings for each nugget for this project.

Check that both ends of the jump ring meet evenly and are closed. If not, use your chain-nose pliers to accomplish this. With one of the metal files, file a notch, groove, or flat spot into the end portion of the nugget where the jump rings will be attached. The size and shape of your nugget and the size of your jump rings will determine how much to file (as well as which file to use). Filing removes any impurities that would prevent the solder from adhering properly.

Secure the nugget in one pair of the self-locking tweezers. The tweezers holding the nugget are then secured in the "third hand," leaving both of your hands free to work on the nugget. Using your second pair of tweezers, pick up the closed jump ring and position it in contact with the filed area where you want to solder it on the nugget.

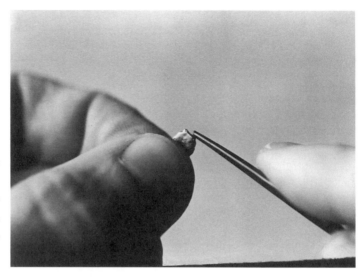

To attach a jump ring, file a flat spot on the end of your nugget.

Place the jump ring so that its opening will be covered by solder. With your other hand pick up your flux brush and dip it into the flux. Apply the flux to the area that you are going to solder. Apply flux also to a solder snippet. Use the flux brush tip to transfer the snippet to the area to be soldered.

Apply the blue part of your torch flame to the **underside** of the joint formed by your nugget and jump ring, keeping the flame more towards the nugget than the jump ring, as this acts to **draw** melting solder into the joint. Never apply heat directly to the solder itself. Always keep the torch in motion while soldering to heat an area evenly. Hot spots in the metal created by holding the flame motionless will result in the solder forming into lumps.

After the jump ring has been soldered to the nugget, remove it from the tweezers which are securing it to the third hand with your second pair of tweezers (it is still hot!) and drop it into your pickle pot.

Pickle works best if it is heated to around 140 degrees Fahrenheit. Pickling pots that keep the pickle at a constant temperature can be purchased from hobby and jewelry supply stores, or you can set your pickle container on an electric heating coil on a low setting. The pickle removes the "fire scale" created by the intense heat of the torch. After the fire scale is removed you will probably want to buff the finished joints with a water soluble red rouge compound. You can do this with either flexible shaft machines, arbors fitted with polish-

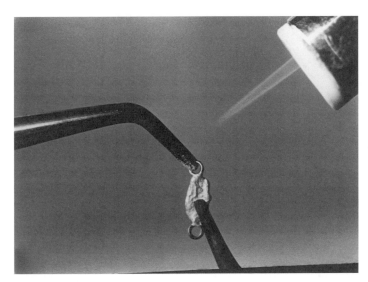

After fluxing the surface to be joined, braze or solder it to the ring.

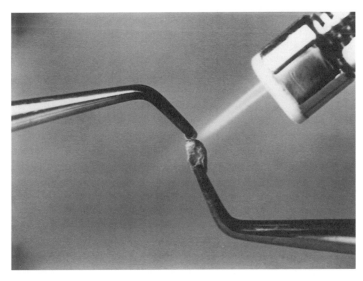

Apply the torch flame to the bottom of the nugget, not to the ring.

Buff to a sheen.

ing buffs, or by applying some of the rouge to a damp towel and hold-ing the nugget and rubbing it vigorously over the rouged towel. Rouge should be sufficient to polish up your joints and nuggets. Stronger polishing compounds are available, but these might smooth the nuggets too much, making them look unnatural.

If you have a small rock tumbler, load it with water, add one to three tablespoons of TSP, or soap powder, and the soldered nuggets. Seal the tumbler and let it run for about 30 minutes. Do not use any of the rock shaping/polishing compounds that normally are used in tumblers. These compounds will grind your gold away into nothing. Remember that gold is very soft and malleable. Thirty minutes of tum-bling together in a cleaning compound such as TSP or soap is all that is necessary.

Wear safety glasses when using either flex or arbor equipment. The speed of the turning shaft will be great enough to flip the item being polished out of your fingers and across the room or directly back at you. For small pieces that are difficult to hold, the rotary tumbler method is your best bet.

When your soldered gold is polished, you can design jewelry by opening and adding one or more jump rings between the jump rings that are soldered to the nuggets, or you can add lengths of chain for a different look. Using your two pairs of chain nose pliers, pry the jump rings open by twisting them sideways. Slip an open end through your soldered jump ring, then twist the open jump ring closed by repeating the sideways twisting motion. You do not need to solder these "filler" jump rings, only those directly attached to your nuggets.

You will want to attach a clasp, or "spring ring" to the ends of your necklace for ease in removal unless your necklace is long enough to slip over your head. If making a hat band you will need to fasten the band in place securely so that it won't fall off.

By using chain, jump rings, and soldered nuggets, you can make an array of stylish and unique jewelry, as depicted in the accompanying illustrations. Remember to use heavier chain on bracelets than on necklaces. Also, small nuggets work best as earrings, larger ones as pendants.

If you shy away from soldering work, you can purchase jewelry "findings" from hobby or jewelry supply outlets. Using a two-part epoxy that bonds metal, you can glue jump rings or bails to nuggets for making simple pendants. I don't recommend this approach because you may end up losing your nugget if the glue-bond fails.

Consider making jewelry from your larger nuggets by wire-wrapping them. Gold wire may be purchased and wrapped around nuggets in an open design which will showcase your nugget while holding it securely. Wire-wrapping methods are popular for creating wearable pendants from crystals and other oddly-shaped objects. A pair of chain-nose pliers, your fingers, and your imagination are the only tools you will need. Keep in mind that you need to fashion a loop of wire through which you can pass a length of chain so that you can wear your pendant. Two publications which may assist you in wire-wrapping your nuggets are two Gem Guides Book Company publications, *Contemporary Wire Wrapped Jewelry,* by Curtis Kenneth Leonard and William A. Kappele, and *Jewelry Making for Beginners — the Scroll Wire Method,* by Edward J. Soukup.

You can create a display showcasing your finds. To do this you'll need a picture frame, complete with backing and glass. Gold looks best against a black, red, or green background, black being my favorite for a very dramatic presentation. A regular picture frame will work well for relatively flat nuggets. For thick nuggets you will need a "shadow box" frame, which is similar to a picture frame but much deeper, and often comes equipped with a hinged, glassed-in front piece for easier opening.

Shadow boxes can be purchased in many shops that sell craft supplies and in most frame shops. Line your picture frame or box with fabric in your choice of color. Using a "tacky" or rubber cement, dab a bit on the back of each nugget and place it on the fabric within the frame in a design you like.

It is best to design your nugget placement before beginning to glue. You may choose to hang this eye-catching display on a wall in your den or private study; if after a time you decide to display your nuggets in another way, or make jewelry from them, you merely rub the glue from the back of the nuggets and your gold is ready for another use. Always remember to remove all grease, glue, and grime from your nuggets if you intend to do any soldering, or if you plan on epoxying nuggets to findings.

If you are lucky enough to find a nice quartz stringer that bears visible gold, remember that gold/quartz — or gold ore sold as specimens — will fetch a far higher price than the usually minimal price of its gold content. Of course, it helps if the gold in your specimen can be actually be seen without too much squinting!

If you have any lapidary (stone cutting and polishing) skills you can create beautiful pieces of jewelry by slabbing chunks of gold ore with a diamond-bladed slab saw. The slabbed pieces are then inscribed with a shape such as an oval, round or rectangle with an aluminum "scribe." Use a diamond-bladed trim saw to remove excess material. Slab and trim saws cut only straight lines, so grinding wheels are used to give the piece its final shape. Successively finer grades of grinding/polishing compounds are used to bevel, shape, and bring up a glossy polish. You end up with a gem of enduring value that can be set as a ring, belt buckle, brooch, earrings, money clip, or whatever.

Chunks of gold ore can be fashioned with Dremel-type carving tools into lovely small sculptures and works of art. When carving quartz-lode containing gold, keep in mind that the gold may tend to adhere to your grinding and polishing tips, necessitating frequent cleaning. Practice your carving skills on common quartz rocks before attempting to carve your valuable piece of gold ore!

Using your own gold flakes and nuggets for jewelry or display, or crafting beautiful cabochons or sculptures from your found gold-in-quartz ore are all wonderful and satisfying ways to increase the value of native gold, as well as to satisfy your innate artistic ability and expressiveness. Creating a thing of beauty is also an extraordinarily satisfying way to indulge the passion that this golden element inspires in us. Designing and creating our own golden treasures allows us to steep ourselves completely in this wonderful, rare and, elusive substance which has hopelessly captured the human fancy — Gold!

Staking a Claim

What should you do if you find a great placer spot, or a gold-laden lode, and you want to stake a claim on it? What rights does staking a claim give you? How do you go about staking a claim?

Staking a claim will only confer upon the claimant the mineral rights to a parcel of land. A claim confers no other rights, unless you are applying for a patent — a subject that will be discussed further on. It all begins with the 1872 Mining Law, the concept of "good faith," and the "prudent person" test.

The 1872 Mining Law states that citizens of the United States, or those who intend to become citizens, have the opportunity to search for, and locate, certain valuable mineral deposits on public lands in the United States. The Law also sets forth standards and requirements for making mineral claims under a variety of circumstances and conditions, which apply to both placer and lode gold deposits, and other "locatable" minerals.

There are those who are presently attempting to overturn the 1872 Mining Law, and that some changes in procedure for making and maintaining claims are likely to occur in very soon. There have been some minor changes already, and more are expected. In 1976, locating claims in National Park-managed lands became more difficult because of the Mining in Parks Act (16 USC 1901), which withdrew all current and future units of the National Park System from mineral entry. Due to this Act the National Park System (NPS) has the authority to regulate mineral development on both patented and unpatented claims. This does not mean that mining is not allowed, only that many more restrictions apply.

The "good faith" requirement assumes that when you stake a claim you are doing so for the purpose of extracting mineral values and not for the purpose of procuring land for residency, storage, dumping, or any purposes other than those related to mineral extraction.

The Bureau of Land Management (BLM), a Division of the U.S. Department of the Interior, manages and regulates the location and filing of mining claims on public land. Here again the "good faith rule" applies. The BLM does not take the time to check if a person's claim is measured out correctly or staked over an existing valid claim. That is the sole responsibility of the claimant. The only time the BLM steps in is if there is some reason to believe a claim is in violation of land-use statutes or in some respect infringes the agency's ability to manage the land. Remember, the BLM does not step in to solve overfiling or "claim jumping" allegations between parties. This is strictly a civil matter and must be handled by the state's civil courts.

Some years ago I staked a lode claim in a National Forest. Unknown to me, until I did further research, was that my claim was staked over a patented site, and was therefore invalid. Do your research, as many patented areas still exist. The holding of a patent confers all rights to the patent holder, including the "privilege" of paying property taxes on the patented parcel.

It is the responsibility of the claimant to locate and mark the claim appropriately. If this is not done properly the claimant may lose his or her rights to someone else who does the paperwork and meets other requirements. Remember, a claimant's rights are only valid if the process is completed accurately.

The BLM does not maintain maps of existing valid mining claims. The accuracy of a claim is the sole responsibility of the claimant. Furthermore, the rights of a claimant may be revoked should errors or falsifications be discovered during either the location or filing process.

It is the responsibility of the prospector seeking to stake a claim to determine if the proposed claim will overlie or infringe existing valid claims. If it does, the claim cannot be made — unless the prospector hopes to locate a lode claim upon an existing placer claim. This must be done by agreement of both claimants, as it would be illegal to enter upon the placer claim for the purpose of staking a lode claim without the placer claimant's permission. The staking of a lode claim within the limits of a placer claim does not provide any rights to the placer gold, or visa versa. The two claims are mutually exclusive unless located by the same individual. Even so, separate filing and fees remain in effect for both claims. If a prospector later locates lode gold upon his placer claim, he must file a Lode Mining Claim Location Notice if he wishes to procure rights to the lode deposit in addition to those rights already procured via his placer claim.

A claim cannot be staked on private property without the property owner's permission. This is something that must be dealt with by negotiating with the property owner. Remember that the 1872 Mining Law pertains to staking claims on **public land.**

Locatable minerals — those subject to either location and staking of claims — form a huge category and consist of almost everything **except** sand, gravel, cinders, stone, pumice, and clay. These items fall under the classification of "salable." There is another classification which is called "leasable," but these two classifications do not concern us here. Gold, whether placer or lode, falls into the "locatable" category and is the category of interest to recreational and professional prospectors alike.

The "prudent person" rule states that, " ... where minerals have been found the evidence is of such a character that a person of **ordinary prudence** would be justified in the further expenditure of labor and means, with a **reasonable likelihood of success** in developing a valuable mine ... "

A person may stake a claim based upon only the knowledge that a locatable mineral exists, or upon evidence that it does. It need not be determined prior to staking a claim that marketable, or profitable, quantities of that mineral exist. One need only suspect that this may be true. It is assumed that a "prudent" person would not continue to work an unprofitable claim. However, there must be **evidence** that the mineral exists in some quantity, or to warrant the filing of a claim.

For placer claims, this "evidence" consists of reasonable quantities of placer gold. On lode claims, "evidence" could be in the form of quartz or other geological indications that a lode may exist. Of course, assays or other tests would have to be performed to verify that gold, in some quantity, is present in order to justify maintaining the claim. It is assumed that the claim would be relinquished by a "prudent person" should tests reveal that not enough values exist to warrant further labor and expense.

A lode is defined as "a valuable lode, vein, ledge, tabular deposit, or other rock in place between definite walls or boundaries." The legal definition of a placer is "all forms of deposit, excepting veins of quartz, or other rock in place," in other words, a disseminated deposit.

Public lands open to location and claiming are restricted to the following 19 states: Alaska, Arizona, Arkansas, California, Colorado, Florida, Idaho, Louisiana, Mississippi, Montana, Nebraska, Nevada, New Mexico, North Dakota, Oregon, South Dakota, Utah, Washington, and Wyoming. Other states allow recreational prospecting in many areas but it is not possible to stake a claim under the provisions of the 1872 Mining Law.

When staking a claim it is necessary to physically mark the boundaries of the claim as well as the "discovery" site — the site where the claimant made his or her initial mineral discovery. The boundaries of the claim are generally the four corners of the claim. If the physical conditions of the site are such that marking the four corners is impossible, then it is permissible to mark them as nearly as possible, but the claimant must include written information detailing the actual boundaries.

Marking the boundaries consists of erecting a wooden post not less than $1\text{-}1/_2$ inches in diameter that projects at least two feet above the ground, or a stone monument that is no less than two feet high. A capped metal post or rod may be used as long as it is greater in diameter than one inch and projects two feet, or more, above ground.

It is common for prospectors to secure a white four-inch-by-four-inch post within a rock monument. The monument is composed of large boulders piled together. Prospectors generally leave a copy of the claim notice in a sealed container within the rock mound. One may also post the copy upon the post — usually sealed in plastic for protection.

In former days prospectors used "Prince Albert" tobacco canisters to protect copies of their claim notices from the weather. These days prospectors use metal Band Aid containers, small plastic bottles, or seal the claim notice in a plastic bag (not as long-lasting as the other two options). Glass jars are not recommended as they are too easily broken.

Requirements for a placer claim

1. Only one discovery monument is required *where there was a U.S. land survey* performed. If a U.S. land survey was performed that area will be covered by a topographic map. Corner markers are not required for placer claims, although the BLM recommends that end monuments be erected at either end of a placer claim — only for the purpose of designating the limits of your claim. This eliminates the possibility of another prospector accidentally overfiling upon part of your claim.

2. Only one claim notice is required to be posted on a claim.

3. A single claim consists of 20 acres. A group of claims held as a unit or "association" must have one party of interest listed for each 20 acres of the unit. Separate location notices and fees are required for each 20-acre claim. Topographic maps help make staking a claim a relatively simple process.

Claim requirements for a lode

1. Claim notices must include the name of the lode. The names of placers or lodes can be descriptive, or personally meaningful in some way, such as these authentic, historic names: The Penelope, The Primrose, The Agamemnon, The Sixteen-To-One (referring to the ratio of silver to gold in coinage of former days), The Ruby Mine, The Young America Mine, The Father's Day Mine. The name may give a clue to the method, or some event associated with the discovery of the claim site. In days past prospectors often named their claims after a beloved mother, aunt, or girlfriend. A partner and I once had a couple of placer claims in the Colorado Desert of California which we had named "Tomorrow's Dreams." Sadly, they didn't amount to much; that area is now part of the Chuckwalla Mountains Wilderness Area.

2. The name, current mailing address, or residence address of the locator of the claim.

3. The number of linear feet that are claimed, along the length or course of the vein, in either direction from the discovery point. A claim shall not extend for more than 1,500 feet along the course of the lode, or vein, or be more than 300 feet on either side (600 feet total width) of the centerline of the vein.

4. The date of location is also considered the date of posting the notice.

5. A description of the type of monuments — both the discovery monument and the corner monuments — and their locations. These monuments must be erected within 60 days of locating the lode. Corner monuments are required for a lode claim.

6. A description of the claim in relationship to either natural objects or erected monuments is necessary as a means of identifying the claim.

7. Upon abandonment of a claim the claimant has 180 days to remove all monuments pertinent to the claim. If this is not done, the claimant is subject to a $50 fine for *each* monument left behind and is additionally liable for any cost incurred by the federal, state, or county agency that removes or has removed, any monument.

Note that your topographic map is divided into squares, each square representing one square mile, or 640 acres. These squares are called "sections;" note that each section has a designated number that can be determined by looking at the center of the section.

Along the left and right margins of the topo map you will see indications such as T. 26 S, as an example. This is the township measurement and in this case means Township 26 South. Each township is usually (but not always) composed of 36 sections.

The measurements across the top and bottom of your topo are called "ranges." They are designated by a letter "R" (designating "Range"), a section number, and another letter designating direction. As an example, "R. 34 E." means that you are looking at Range 34 East. All range numbers east of a north-south meridian are designated by "E," or east. If the range number is west of a meridian, you will see a range number with a "W," or west designation, for example, "Range 34 West," read as R. 34. W.

Ranges and townships are numbered consecutively east or west of a north-south principle meridian of longitude and north or south of an east-west base line of latitude where these two intersect. Each range, or north-south strip is six miles in length. While all this may sound relatively complex it will be very clear when you look at a topo map and see how simply it is laid out in front of you.

To identify your claim you need to purchase a topographic map which depicts the particular quadrangle in which the site of your proposed claim is located. Identify which section of the quadrangle contains your proposed claim. That section will be identified by its own number. Determine the range and township of your section by determining which township and range the section lies in on your topo map.

Thus, your claim may be in Section 30, Township 6 South, Range 15 East of the San Bernardino Baseline Meridian. However that only tells what section your claim is in. A section consists of 640 acres and a single placer claim is considered to be only 20 acres. So, this must be further broken down in order to give a precise coordinate for your claim.

Each section of 640 acres is broken into equal quarters consisting of 160 acres each. This breakdown is not indicated upon your map. You do this yourself by dividing the section into four equal squares that represent 160 acres each. The top left square is the NW, the top right square is the NE. The bottom left square called the SW and the bottom right square is called the SE. Each of these squares is further broken into four more squares of 40 acres each — we are now starting to get close to placer-claim size! These 40-acre squares, or quarters within quarters are further broken down into quarters of 10 acres each. Remember that each quarter, or quarter of a quarter, no matter how large or small, each has a NW, a NE, a SW, and a SE.

It is these 10-acre sections that you can use to designate your 20-acre claim. Now that you have narrowed down the quarters in which your claim sits, the erecting of monuments to delineate the precise locations of the discovery point and the corners (in the case of a lode claim) is all that remains of the physical designating of your claim.

It gets a little more interesting when your claim sits across 10 acres of the SW and 10 acres of the SE, or laps over on even one or two more quarters. Rarely are claims tidily located in one quarter of a quarter of one section. Placer Location Notices have areas that you check to describe the quarter by quarter breakdown of your claim.

Your claim may even lie across Section lines. A partner and I once had two 20-acre claims, end to end, which cut diagonally across the area where corners of four sections met. The actual physical marking of the claim (erecting of our discovery and end monuments) was no problem, but trying to describe it on paper was a little confusing. However, by taking it section by section we managed to accurately describe the location of our claims. Our claim — to add even more complexity to an already complex situation — also lay between three patented claims and was somewhat of a nightmare to physically plot so as to not overfile on the patents, thereby rendering our claims invalid.

The written description of your claim will include all of the quarter descriptions down to 10 acres. As an example you could have a claim that is located in the SW quarter of the NW quarter of the SW quarter of Section 30.

Once you have your claim coordinates figured out it is time to fill out your claim form. These are available through the BLM or at most stationery stores in the same files that other legal forms are found, such as deeds. Many prospecting supply stores carry claim forms. The first four forms will be the ones most used by recreational gold prospectors.

The forms available are as follows:

1. Lode Mining Claim Location Notice.
2. Placer Mining Claim Location Notice.
3. Affidavit of Assessment Work.
4. Maintenance Fee Payment Waiver Certification.
5. Mill Site Location Notice.
6. Tunnel Site Location Notice.
7. Notice of Intention to Apply for Mineral Patent.

The word "location" in the context of "Location Notice" means that a person is in the process of performing the necessary tasks to acquire the mineral rights to a parcel of land, such as determining the coordinates, researching whether the land is open to mineral entry or under previous valid claim, erecting monuments, posting the location notices, etc.

Once the initial research and posting of the claim is complete, you must file the claim. After verifying the accuracy of the physical description of the site location and that the area is open to mineral entry, the prospector must have the Location Notice recorded by the

County Recorder of the county wherein the claim is located. In addition, the claim must also be filed with the state office of the BLM. Both filings must be recorded by both the county recorder's office and the BLM state office within 90 days of location. It is highly recommended that the filings to both offices be sent by certified mail so that the claimant has proof that he or she attempted to comply with the regulations.

Under the 1872 Mining Law the provisions for the 19 states where it is legal to locate mining claims are largely uniform, but it is likely that there are some minor variations from state to state; it is always to your advantage to be sure of both state and country regulations when filing a claim. This is especially important at the present time when various factions are whittling away at the 1872 Mining Law from county, state, and federal levels. The information given here applies, at the present time, to the State of California. However, the federal Department of the Interior regulates mining in the 19 states, and California is basically representative of what you will encounter regardless of the state in which you are filing a claim. In the chapter, **"Publications and Organizations to Assist You"** I have included the address and phone number of the office of the Department of the Interior that serves each state.

Various filing fees are currently in effect; these fees are subject to change. I will give an example of what you may expect to pay when you stake a claim, maintain, or transfer a claim. Not all of these fees and filings will be applicable all the time. Some of them will not pertain to the recreational prospector. You need to be aware of them nonetheless, as each event pertaining to your claim will have a fee and filing requirement.

1. New Location Service Fee — $10.

2. Location Fee per claim — $25.

3. Assessment Year Maintenance Fee — $100. This fee may be exempted under the "Small Miner Exemption," which allows a miner holding less than 10 total claims (nationwide) to perform maintenance work on each claim **worth** $100, rather than paying $100. If a prospector uses the Miner's Exemption he or she must file the Maintenance Fee Payment Waiver Certification. There is a $5 filing fee for each claim. The Maintenance Fee Payment Waiver must be filed with the BLM state office where your original location notice is on file.

Where several claims are located as a unit, one may do the $100 **worth** of assessment work on any one of the claims as long as all claims in the unit benefit from the work. Maintaining a road that accesses all the claims would be an example of valid assessment work. Under "good faith" the work done should be obvious and of practical benefit to the claim or claims. Save any receipts for services incurred on the improvements made to your claim. If $100 **worth** of assessment work is not done under the "Small Miner's Waiver," or you have more than 10 claims, $100 must be paid for each claim held. This fee must be paid prior to August 31 of each year in order to hold the claim for the assessment year, which starts on September 1 and runs to September 1 of the following year.

This present designation of an "assessment year" is valid until September 1, 1999. Changes may occur after that date; be sure that you contact the BLM to determine if they will and what they may be. Failure to keep up on current claim requirements may cause your claim to be invalidated. Within the assessment year that claims are filed no assessment work or payment of assessment fees is required.

4. **Amendments/Transfers of Ownership per claim** — $5. This form is used to correct an erroneously recorded location site, if the owner wishes to change the name of a claim, or has discovered that other recorded information pertaining to the claim is wrong. This form must be filed with both the county recorder and the BLM.

5. **Affidavit of Annual Assessment per claim** — $5. This form is completed in order to detail the type of assessment work that was performed for a claim, such as road work, rebuilding of fallen monuments, repainting the posts, etc. It is advisable that a miner take both before- and after- photos of the assessment work. If a question should arise concerning the validity of your claim, you will have acceptable proof that you did perform the stated assessment work. The "good faith" requirement sometimes needs some physical evidence to back up your assessment claims.

6. **Notice of Intent to Hold per claim** — $5.

7. **Petition for Deferment of Assessment Work** — $5.

8. **Mineral Patent Application** — $250. The first application is $250. A filing for each additional patent is $50.

9. **Mineral Survey Application** — $950. Each additional site is $375.

Numbers "8" and "9" are of little concern for the recreational prospector.

The BLM issues a serial number that pertains to your claim upon receipt of your initial filing of a claim location notice. That number is to be used on all future correspondence and filings that you submit in reference to a particular claim. You can identify your claim's serial number as it will be prefixed by "CAMC."

The transfer of ownership of a claim will also need to be recorded via a standard "quitclaim deed," the type commonly used to transfer any type of real estate. The transfer of ownership (quitclaim) must be filed with both the country recorder and the BLM. Include the claim's serial number and name, and the $5 filing fee. Record and send within 60 days.

Often, individuals locate and file claims together. Should one of these individuals die, the other will need to notify the BLM and provide a copy of the will in which the deceased relinquishes to the survivors his or her interest in the claim (or court decree where no will exists), and a copy of the death certificate. If the deceased is a spouse to the surviving claimant the agency will accept in lieu of the above a notarized statement that the survivor is now sole owner of the claim. There is a $5 fee for filing.

Following is a listing of the types of assessment work that the BLM finds acceptable (not all of these may pertain to recreational gold prospectors):

1. Construction and maintenance of roads to the claim or claims.

2. Construction and maintenance of buildings pertinent to the claim.

3. Installation of equipment that assists with the extraction of ore or placers.

4. The purchase of supplies that assist in extracting the ore, such as blasting materials or other such material essential to the extraction of values.

5. Underground structures that pertain to the mine (should a recreational prospector turn professional).

6. The hiring or installation of security persons or devices.

7. Removal of tailings that result from extraction, or the overburden overlying the values, but only for the current year.

8. Mineral surveys or assays.

Below is a list of work that does not meet the assessment requirements:

1. Buildings that do not serve the claim.

2. Prospecting or exploration for discovery, rather than for extraction.

3. Any services that are not necessary to maintain the claim.

4. Supplies that are purchased but not used.

5. Stockpiled values from a previous year.

About the only time a filing fee is not charged is when a claim is abandoned. In the case of abandonment, a letter to the country recorder and the BLM agency should be submitted. The Maintenance Fee Payment Waiver has a box to check if you desire to abandon any claims. Remember to remove your claim monuments when abandoning a claim. Fees may be charged if you do not.

The question is often asked, whether or not it is permitted for the claimant to live upon his or her claim. It is permitted, but there are some restrictions that apply. The claimant must be actively engaged in working the claim in order to reside upon it. Residency must be predicated upon the regular operations of prospecting, exploring, extracting and processing on an ongoing basis. This means that mine operations must not be restricted to weekends and must essentially be one's primary occupation (not usually a factor for the recreation prospector, although not unheard of).

While the residency entails that the primary occupation be the prospecting for and extraction of values, one does not have to reside full-time at the claim. Residency may be either full-time or part-time as long as mining is the primary occupation.

A word of caution: miners must contact the BLM or U.S. Forest Service (if the claim is on Forest Service-managed land) prior to putting up a residence on a claim. There are many restrictions upon construction of buildings on claims; a final determination will be made by the land agencies. In many areas the Forest Service does not allow any new construction but will allow maintenance of existing buildings. Don't overlook this point, or you may find yourself receiving an order to vacate and remove your structure, not to mention a variety of fines. Should you be able to gain permission to erect a building, it must be constructed in accordance with state and county codes.

The subject of mining patents often arises. There are many old patented mines. A patent is a document that conveys title, not just temporary use, to mineral values as well as to surface features.

Old patents in National Forests may be regarded as "a fly in the ointment" of the bureaucracy. Patent holders are often contacted by the Forest Service and offered the opportunity to trade the patent for land elsewhere. As an example, there is patented land on which an old gold mine sits in the National Forest near my home. The location and discovery of the mineral values of this old mine took place in 1897; in 1929 a patent, signed by President Herbert Hoover, was issued on the property. The current owners of the patent contacted me to survey their property for remaining gold values. I took both mill site samples and chip samples from the vein within the mine, for both assay and spectrographic analysis. During that time I was shown several letters that the patent holders had received from the Forest Service, offering to make a land-trade for the patented property. The Forest Service was anxious to get hold of the patent for the water shed and timber rights. There wasn't much in the way of timber, as much of the patented area was above the tree line, but a small, year-round creek did run through the property.

While a citizen still has a right to apply for a patent, the issuance of a patent is another matter entirely. It could well be an unwritten rule that patents are simply not issued these days. A book of mining regulations recently issued by the BLM states, "The patenting process can be complex, expensive, and lengthy." Take this to heart and don't expect much more than frustration should you decide to apply for a patent. The complexity of the process is almost sure to result in some errors, any of which are grounds to deny the application.

The information provided in this chapter covers the basics of filing a mining claim. Particular questions on areas not covered here should be addressed to the BLM, or your state's office of the Department of the Interior.

The BLM puts out a large book entitled *Location and Patenting of Mining Claims and Mill Sites in California*. It is packed with all the information you need for staking claims in California. Ordering information for this book is included in the chapter, "**Publications and Organizations To Assist You.**"

We prospectors often expect that BLM and Forest Service employees and officers exist only to make our lives more miserable, and certainly there are some horror stories to tell. Although some may have had contrary run-ins with agency members, I have to state that all of my contacts have been positive. I hope that it will remain so.

I firmly believe that it is how we approach people that most often affects how those contacts proceed. I suggest that you do everything in your power to obey the regulations and requirements concerning the location and filing of claims. There really is a lot of latitude within the current regulations. By being honest, forthright, and reasonable you will generally not create opposition.

On the other hand, if you try to pull a fast one and operate in a manner inconsistent with the requirements, I can guarantee that you will eventually run into trouble. Conduct yourself in an honorable and reasonable manner and you will benefit from the assistance, knowledge, and advice of these officers. Most are very familiar with the details and history of the tracts of land that they patrol and can often help you locate something you are looking for.

If you get into trouble out in remote areas land-agency officers are likely to be the ones who will find and rescue you. They can help the prospector in a variety of ways, and we can help them by following the regulations. Keeping informed of any changes in the regulations benefits both parties, thereby making our encounters mutually pleasant and beneficial.

More about Lode Gold

Most recreational prospectors find concentrations of placer gold, but occasionally we choose to search for and locate lode deposits. If you prospect for pocket gold you are very likely to find lode gold; many of us cannot resist the lure of legendary — and lost — ledges of quartz, rumored to be shot through and through with golden stringers.

There are many such tales of lost mines and veins of gold. These are fun to look for, and I find that sometimes I cannot resist the adventure of looking for a lost lode. As always, the best way to find gold is to search for it where it has already been found.

The best method for actually locating lode gold is to research old lode mines and mill sites. Use your detector to go over tailings piles. If you receive no signals, use your shovel or pick to scrape down eight inches or so to the next level, and run the detector over that.

Check old mill sites for un-milled ore that could contain gold. The detector will "see" what may have been hidden from view to long-ago miners. A piece of ore with no visible gold on its surface may hold a golden surprise inside! If you get a signal on a piece of ore, use your pick to crack open the piece for a look, or take it home and slice it on a diamond-bladed slab saw. Just remember that you may also get signals from hot rocks in many gold-rich areas. There is a discernible difference in sound between hot rock-signals and those of gold. If you use a detector regularly you will learn to distinguish between the sweet song of gold and the dull "thud" that signals a hot rock.

Areas surrounding old mine workings are good places to begin the search for lode gold.

If you prospect for pocket gold, you will be finding lode gold that has eroded out of decomposing quartz veins. If you have a yen to search for lode gold, or just happen to follow your detector signals to a vein exposure, you will need to know a bit about prospecting for and recognizing the rock-types that exhibit the potential for carrying gold.

As you may recall from the chapter "Electronic Gold Prospecting", lode gold, as well as its rock matrix, may be stained with iron oxides. These oxides are referred to by several names, such as "gossan," "bloom," or "iron hat." This staining occurs when soluble minerals are dissolved and carried by groundwater or moisture, entering the fine cracks and fissures that occur in rock. Various minerals create the color, or colors, of the oxides occurring at a site. Some of the most common colors of gossan, and their mineral associations are:

Rust red, yellow, black, or brown = iron or chalcopyrite
Blue or green = copper
Red or pink = cobalt
Pale green = nickel
Pale yellow = molybdenum
Black = manganese
Bright colors of green, yellow, or orange = uranium
Red-orange = mercury

Quartz outcroppings like this are good prospects for lode gold.

While gossan can be an indicator, it can also indicate very little, particularly in the case of copper, mercury, or manganese. The axiom, "It takes only a nickel's worth of copper to turn a mountain green" is based in much wisdom. While pale green may indicate the presence of nickel, it is also the color of one of the most common "rock-forming" minerals, the "peridotites."

Your best gossan indicators are those associated with iron, particularly of the yellow-brown coloration or rusty-red. Rust-colored rocks have long been an indication that gold may be in the area. Although pyrite is often called "Fool's gold," its presence in vein material is an indication that gold may also be present.

The prospector must be able to distinguish oxide-stained rocks from rocks that naturally occur in these colors. Gossan appears as staining that covers and interpenetrates rocks of another color—almost like a rust stain, but often more glossy and smooth. The term "iron hat" alludes to the fact that gossan is something either "worn" by or "added" to a rock specimen. Oxide stains usually cannot be rubbed off, but they can be dissolved away.

The prospector must be aware that in many areas lichens grow on rocks, and could be mistaken for gossan. Desert areas are rich in lichens that mimic the colors of gossan. Lichens can usually be scraped off, while gossan cannot. Lichens usually have an "un-rock-like" texture, almost fibrous or "twiggy."

Besides lichens, deserts offer another problem, called "desert varnish." Desert varnish is a manganese or iron oxide coating of black or brown that covers long-exposed rocks and outcroppings. The true nature of these rocks can only be seen by chipping the sample with a rock pick, which reveals the actual color of the rock in question. When prospecting for lode gold I always carry a rock hammer in my prospector's day pack. A sharp blow to the edge of a specimen will chip off a flake so that the fresh, unexposed interior of the rock can be seen and appraised.

The author on a lode prospecting trip in Arizona.

Prospectors locate lode gold by following "float" (eroded pieces of country rock) back to the source, as was discussed in the chapter, **"Electronic Gold Prospecting."** Float looks promising if it has a gossan coating or has empty cavities or cells, called "box work." Box work indicates where gold or gem-crystals may have once been embedded. This is a valuable clue when prospecting for pocket gold.

Lode gold specimens may contain tiny specks of gold that are invisible to the unaided eye. To determine if gold is present in this case, crush the specimens with an iron mortar and pestle, then pan out the resulting sand and dust. At the end of the panning process the gold in the specimen will be evident. You may need a loupe to view very fine or flour gold. It is sometimes necessary to use mercury to process, or amalgamate, flour gold from crushed specimens. Retorts for safe mercury-processing are sold by most mining supply stores, as are booklets detailing safe and environmentally proper uses of mercury.

You can also hand-crush specimens to determine if they contain gold, although hand-crushing is not efficient for processing an abundance of ore in order to recover gold. Recreational gold prospectors who take up lode prospecting may want to purchase or build a small-scale crusher that will fit on top of a five-gallon plastic bucket. The bucket collects the crushed ore for later panning or processing with a mechanical processing unit. These processing units are a great time saver for the prospector who can't sit around for hours panning gold from crushed ore, or prefers not to use mercury. A variety of "automatic panners" on the market will process your fines quickly and efficiently while you go about your other duties. They are manufactured in "screw," "bowl," "double-wheel," and "mini-sluice-type" models and may cost $35 for a small pump-operated bowl-type or $3,000 for a model that processes two yards of material in less than a minute.

A small ore crusher costs around $750 new, and weighs between 80 and 90 pounds. Small crushers are a useful investment for the recreational prospector who enjoys prospecting for lode gold. Shop around for a used crusher. Be sure to check the classified sections of prospecting magazines.

Larger crushers can cost $5,000 or more, which effectively places them out of the realm of the average recreational gold prospector. A woman prospector I know built herself a small crusher after careful examination of a model that she saw while shopping at a major distributorship of recreational prospecting supplies. On a subsequent prospecting trip, we tested the crusher by processing ore samples from some ore-piles we had located near a group of abandoned gold mines.

On its first trip through the crusher the ore was reduced to pea-sized gravel. A second spin through the crusher resulted in sand that was suitable for panning and from which we recovered some gold. The crusher cost around $60 to build and worked great. Building your own crusher would be a money-saving do-it-yourself project. Many abandoned mines have on-site piles of ore that were never processed just waiting to be explored with a metal detector or crushed and panned.

Contact zones, where two dissimilar rock-types meet, are also good places to prospect for lode gold. In addition to gold, there is a tendency for crystal pockets of worthwhile and valuable gems to form in contact areas,

Several years ago a couple of my prospector friends were exploring a canyon in the Chuckwalla Mountains of California. This canyon had a history of lode and placer mining during the 1930s. While driving through the canyon they spotted a contact zone in the canyon wall and discovered a sizable pocket of fine "grossular" and "hessonite" crystals in the red and green rodingite contact, which ran through the gray country rock of the canyon. Grossular and hessonite are two types of garnets that are prized for use in jewelry and thus were a profitable find. After several subsequent trips, during which we collected all that we wanted, the location of the find was published so that other prospectors might discover and collect these garnets. A savvy prospector is open to all opportunities, not just the golden ones!

Gold often occurs in silica-rich rock such as granite, schist, or gneiss, as well as in quartz. Any field guide to rocks and minerals, which is a useful companion for any prospector, will illustrate and describe these rocks. One of my favorite field guides is *A Guide to Field Identification — Minerals of the World,* by Charles Sorrell, Golden Press/Western Publishing Co., Racine, Wisconsin: 1973. This book is packed with useful information and is very well illustrated. There are no photographs of minerals. Instead, very detailed color illustrations represent the various minerals. In many ways the illustrations have proven more useful than photographs. This is the book I refer to most often.

Another often-used favorite is *Simon & Schuster's Guide to Rocks and Minerals,* edited by Martin Prinz, George Harlow, and Joseph Peters (New York: Simon & Schuster, 1977). This book contains great photographs that clearly depict rocks and minerals in the forms in which a prospector is likely to encounter them in the field.

Mineral kits can greatly assist the hard rock prospector to become familiar with mineral types.

Many so-called field guides are filled with exceptional photos of rare, museum-grade specimens that are nothing like what one would encounter on a prospecting trip. The prized specimens featured in some of these books come from deep within the earth of privately owned mines — mines that are off limits to the average collector. These rare and valuable specimens are sold for great sums of money on the collector's market to wealthy individuals and museums.

The prospector needs a field guide that will help him or her to identify the types of rocks that are common indicators of their type, rather than unique specimens. It should also give the chemical composition of the rock-types. In many cases this information will help determine how your specimens should be cleaned, as well as the likelihood that gold could be present in a particular geologic area, based on the mineral content of the rocks you've found there.

An understanding of geology and rock-types is not essential for the successful prospecting of placer gold, but it is useful if you have a yen to add lode prospecting to your recreational gold prospecting repertoire. In addition to field guides, mineral identification kits, consisting of samples of rocks and minerals, are a good learning tool to enhance your understanding of geology for the purpose of locating and recovering lode gold.

As mentioned in a previous chapter, geology departments at universities are good places to learn where to mail-order mineral identification kits. The BLM (Bureau of Land Management) offers two kits for sale through its catalog (see the chapter, "**Publications and Organizations To Assist You**"). You can also mail-order mineral identification kits from **Burminco Rocks and Minerals, 128 S. Encinitas, Monrovia, CA, 91016, (626) 358-4478.** This company markets a wide range of kits for use by collectors and schools, so send for their catalog. Of most interest to gold prospectors would be the "igneous," the "metamorphic," "Basic Learner's Kit" and the "rock-forming," or the "ore" mineral kits.

Your loupe or small magnifier is useful for observing indicative crystal structures, which will assist you to identify the various types of minerals you encounter. A microscope is even more useful for this purpose. Too, your loupe will be essential for scanning for those small particles of gold that may not be visible to the unaided eye within your ore specimens.

Remember that placer gold is merely gold that has eroded out of its rock matrix. What we refer to as "placer" was once lode gold. Finding lode gold is often merely a matter of tracing placer gold back to its source, as when prospecting for pocket gold. It is useful to remember, however, that most placer gold has traveled a great distance from its original source. Generally, only in the case of pocket gold is it possible to trace most placer gold back to an original source. Desert placers may sometimes be an exception to this rule, due to intermittent water movement.

If you plan to display or sell any lode gold specimens you find, you will probably want to get rid of any oxides coating the specimens. While black oxides sometimes enhance the look of natural gold — as long as they do not obscure the gold — red or yellow oxides tend to detract from the beauty of a specimen. Gold itself is resistant to penetration by oxides as a rule. You may, however, encounter gold that may be covered up by oxides in or on a specimen. This will be particularly true of ore from a desert environment, where it has been exposed for a long period of time and therefore may be coated with desert varnish. There are many methods for dissolving the oxide staining. Most of these oxide cleanup methods use some type of acid.

Oxalic acid is probably the easiest and safest to work with and is usually purchased as a granular powder from paint stores, where it is sold for bleaching wood. Oxalic acid can be also found in the "painting" section of most hardware stores. It is the same type of acid that is found naturally in many green vegetables, but is very highly concentrated. Oxalic acid will dissolve iron-oxide stains on almost all minerals.

To mix a batch of oxalic acid for present and future use you will need to fill a one-gallon plastic bucket (the type that comes with a lid) $1/_2$- to $2/_3$-full of hot tap water. To this add the contents of a one-pound package of oxalic acid. Stir for about five minutes until the acid crystals are dissolved. Be careful not to inhale any of the oxalic dust. Now add enough water to fill the bucket within one or two inches of the rim. Label your bucket and keep it in a safe place away from kids and pets. While mixing your solution you should wear safety goggles and rubber gloves.

To clean your lode specimens, place them in a plastic container that has enough room so that when you add oxalic solution your specimens will be covered completely in the liquid. Add the oxalic solution and set it in a safe, secure place for several days. Once the stains have dissolved you can remove your specimens. Wear rubber gloves to do this. Place your specimens in a bucket of clean water and let them soak for 24 hours. Rinse thoroughly, and soak another 24 hours. Repeat for several more days, then rinse in running water for several hours and dry.

A hot (110° Fahrenheit) solution will dissolve oxides faster than a cold solution. A warming plate can be used under your specimen container. Never boil the solution, as the fumes are toxic. Excessive heat will also exhaust your solution faster. Normally, the solution can be reused numerous times and will darken with each use. Once it gets very dark and begins to lose its ability to dissolve oxides it should be discarded in an environmentally-friendly fashion, according to the regulations where you live. Most cities have places where you can drop off toxic chemicals for proper disposal. Always wear rubber gloves when handling the acid, as it will be absorbed through your skin and cause harm internally. For handling strong acids, you can purchase special "acid" gloves. The best place to use oxalic acid is a well-ventilated spot in the garage.

If you wish to remove calcite or calcium carbonate coatings from specimens, use hydrochloric acid. Calcium carbonate is a result of the percolation of water. If you have ever been in a mine you will have seen diamond-like sparkles shimmering in your lamp. These particles are carbonates. They build up on specimens that are subject to the constant dripping of water.

Hydrochloric acid will remove oxide stains that are too stubborn for oxalic acid. If one of the components of your specimen is calcium, do not clean it in hydrochloric acid or your specimen will be damaged. As an example, my partners and I would not have used hydrochloric acid to clean our grossular specimens in their rodingite matrix. Rodingite is composed of calcium and silica, and a hydrochloric acid bath would not have been beneficial to the matrix in which the specimens were embedded.

Hydrochloric acid is stronger than oxalic acid and more dangerous. It is sold under the name "muriatic acid" and is used commercially for cleaning and removing rust stains from masonry. It is found at paint, hardware, and pool-supply stores. It is usually sold in one-gallon containers. When using hydrochloric acid, wear gloves and long-sleeved shirts. For prolonged exposure to hydrochloric acid, you should consider purchasing "acid" gloves. Use eye-protective goggles and avoid breathing the fumes — be sure the area you are using it in is well ventilated.

Before using hydrochloric acid be sure to rinse your specimens of all dirt. Dirt is high in iron. Why waste your acid dissolving the iron in dirt? Allow your specimens to dry after the rinse, so the water does not dilute the acid. Place your specimens in a plastic container with some extra room to accommodate the bubbling and fizzing of the acid. Your container should have a lid that you can seal after the addition of the acid. Be sure that the lid has a small hole so that gases can escape without popping the lid. Carefully pour the acid over the specimens so that you do not splash any acid on yourself. After the initial fizzing settles some, put the lid on the container and put it in a safe spot to allow the acid to work on the oxides.

Calcium carbonate and calcite coatings will be removed within a few minutes, so check the progress frequently until the desired effect is achieved. Oxide stains can take up to a week to remove, depending on the amount of staining.

When using acid it is very important to *never add water to acid.* If you do this you will have an explosion of acid particles. You may dilute acid by **adding acid** a little at a time **to water.**

After your specimens are ready you should rinse them in water briefly, then add them to a bucket of warm water to which you have added half a small box of baking soda. Let the specimens soak for three times longer in the soda solution than the time spent in the acid bath. The reason for this is that rock is not solid but is actually quite permeable. The acid that was used for soaking has penetrated into the rock. Change the soda solution each day of the soaking period. If you acid-soaked a group of specimens for 24 hours, soda-soak them for 72 hours, changing to a fresh soda solution each day. Remove, rinse, and dry your specimens. If you acid-soaked for a week, you will need to soda-soak for three weeks, changing the soda solution every few days.

The acid can be reused; as it begins to lose its power it will turn a yellow-green color. Often, specimens soaked in a yellow solution will come out of the acid-soak stained this color. Should this happen merely repeat the cleaning procedure with fresh acid. The yellow stain will dissolve quite quickly.

To discard exhausted acid, add baking soda until it ceases fizzing. At this point it is safe to dispose of the acid. Any time you use caustic chemicals or acids be sure that you dispose of them safely and in accordance with local requirements.

It is also possible to use mechanical means to clean specimens. Sand blasting, steel picks, and brushes are all methods that will work to some degree. Steel picks and metal brushes are recommended for cleaning specimens such as fossils or water-soluble minerals (not generally gold-bearing, although you may have other reasons to bring them home), but if you are cleaning iron oxides, sand blasting is likely to be the only mechanical method that will work.

Beautiful specimens of gold-laced quartz come from Idaho. California is known for gold stringers on snow-white quartz. However, not all lode gold found in the state is of this variety. In northern California, especially in Sierra County, this type of lode is fairly common.

In southern California the lodes are often heavy with gossan, particularly in the desert regions. Whether or not you remove the oxides is a matter of personal choice; acids are the most efficient method for doing so. In all cases when working with acids it is essential to use caution and keep them out of reach of children and pets. Place acids where they cannot be easily knocked over or spilled. If you store acid in glass be sure that the containers are secure in the event of an earthquake. Who needs the double danger of both spilled acid **and** broken glass!

A discussion on lode gold would be incomplete without addressing the problem of entering old lode mines. I highly recommend against it although I fully understand the draw of these old mines. When I first started recreational prospecting in 1970, I was drawn to explore old mines and did so very cautiously. However, after many years and a few close calls, I concluded that venturing into old mines is a very dangerous practice. Now, I only venture to take a peek into the opening. Merely by peering inside, I can still get a feeling for the mine, see what the rock-type is, and gather any specimens that are available just inside the opening. After some hair-raising experiences I prefer prospecting in open and hope that you will, too.

Once, while prospecting in an abandoned mine in the San Gabriel Mountains, a part of the ceiling collapsed, sending a huge slab of rock plummeting down just five feet from where I was working. On another occasion, exploring a mine near Death Valley with friends, we entered a passage that we wanted to check out. Before long we began to smell something like chlorine bleach permeating the tunnel. We bailed out in a hurry. Many mine gases are odorless and you may not realize that you are affected until the moment you black out.

The cool recesses of desert mines are favored hangouts of rattlesnakes, scorpions, and other "nasties." Vertical shafts are one of the dangers that can be stumbled into in a mine. Old ladders and cables may appear to be in good shape but dissolve at a touch, due to rust and age. Timbers and lagging (mine supports) may be termite-ridden, turning to dust with exposure to air.

Despite the many dangers of entering old mines, an old mine in the San Gabriel National Forest once saved my life. This mine is located next to a steep talus slope where rock debris eroding from the mountain flank slips and slides to the canyon below. Over many decades the debris from the slope had filled in the mine adit. Friends and I had located and reopened the adit of this once-lost mine. On one occasion I had hiked up to the mine by myself to take pictures for an article I was preparing on the history and my rediscovery of the mine.

Just as I reached the mine adit to take some photos, an earthquake hit (not unusual in southern California). Rocks began rolling down the talus slope and dust began to rise in huge clouds all around. The sound was like thunder. I was standing on a shelf about two feet wide in front of the mine. The mountain flank fell steeply away to the canyon floor 200 feet below the shelf I was on. The prospects for making a safe run for it were not good. I was dangerously exposed to the talus debris bearing down on me. With no where else to go, I dove into the mine and waited out the rock fall in safety.

A tunnel in the earth is not a bad place to be during an earthquake, as the earth is fairly elastic and a tunnel will move with the seismic waves — unlike buildings, which sway in opposition to the waves. Open country is always the safest place to ride out an earthquake, but in this instance I was surrounded by steep rock-strewn mountainsides and the shelter of the tunnel was my best hope. Aside from this one event, most of my encounters with old mines have ranged from uneventful to downright life-threatening. After many years of disappointment in exploring old mines (as well as a few close calls) I have concluded that abandoned mines are abandoned because nothing of value remains.

The recreational prospector can do very well recovering specimens of lode gold by combing tailings piles and mill sites with a metal detector or prospecting for pocket gold, which happily results in the location of **exposed** lode veins. This is much preferable to risking life and limb in old mines. Too, experience has shown me that more gold — both placer and lode — is to be found in safety aboveground.

Once in a while a recreational prospector will stumble across a really rich find. The only problem is that developing a lode — as opposed to working a placer — is an expensive venture. The lure of gold is so tenacious that before long you will be willing to mortgage your home to fund your mine. It won't stop there!

Rich pieces of ore may lie near the ore chutes of old gold mines.

You could ask friends and relatives to invest in the lode, but the chances are that you would lose your friends along with their money, and your relatives would cease speaking to you. More lode mines fail than succeed. Mining companies succeed where individuals fail because they can afford explore and develop a number of properties. They also have a large pool of investors and very good gold properties.

Another danger of attempting to develop a lode is that prospectors are not truthful with themselves. We tend to take the choicest high-grade specimens in for assay, conveniently ignoring the lower-grade specimens. Naturally, the assay will come back appearing to be quite rich. If a mining geologist were to take samples, he or she would take samples all across the vein, including the low-grade areas, resulting in an assay that would reflect an average percentage of gold overall.

Gold ore samples ready to be conveyed to an assayer in sample bags.

Mark Twain's observation that a mine is "a hole in the ground with a liar standing over it" may reflect the truth that the miner is prone most often to lie to himself concerning the richness of his lode. The temptation is strong to delude ourselves as to what we have found and act accordingly. Before you know it the home, the car, the savings, the retirement funds, have all been poured into a mine that will likely amount to nothing.

Hunting for the occasional lode specimen is fun and can be profitable. Sinking your life-savings into a mine is a sure recipe for disaster. What I recommend to the recreational prospector who thinks he or she has found a rich lode would be to go ahead in locating and claiming it. Then have assays done. You can take the samples yourself, but don't confine your sample taking to the riches parts of the lode.

Having an assay done is not expensive and will tell you whether to proceed or not. If the assays are good, advertise the property for lease to several mining companies, sending along copies of the assays. You may generate some interest and possibly find that a company is willing to send over their own people to do some testing. The result may be the subsequent leasing of your claim for further development by a mining company. You will end up making money to fund your recreational prospecting, rather than losing money trying to develop a mine on your own.

On the other hand you may not find anyone who wants to develop your property. Lode mining is a very "iffy" proposition. Too, in many states mining companies must meet very strict environmental guidelines, which call for them to submit a plan of operation including a commitment to return the area to its pristine state after mining is completed. Many companies will not develop new areas in some of the western states because of the costs involved. A word of warning here would be that if you intend to try to lease your claim to a mining company, do your research, location, and claiming as accurately as you can. No mining company will want to waste time, money, and effort on a project that has not been properly located and claimed, or on someone who has selectively assayed in an effort to make the ore appear more "high-grade" than it is.

If all else fails, mine the lode for the recreation and gold that you can get on your own with a pick, small crusher, and gold pan. If your lode has some really dandy specimens, clean them up and sell them. Just try not to follow up on the notion to develop your lode into a working mine.

If you get really serious about lode prospecting you may want to purchase "geologic maps." These indicate the various rock-types which form land-areas. They are useful for prospectors as well as gem and fossil hunters. To use these maps it is necessary to study a bit of geology so that you understand which rock-types are likely to produce the golden-object of your search. The various rock-types are indicated by colors printed over standard topographic maps. Not only do you get the topographic benefits from these maps, you know precisely what type of rock lies beneath your feet.

Geologic maps are available from: **California Division of Mines and Geology, P.O. Box 2980, Sacramento, CA 95812-2980 (916) 445-5716** or by contacting your state's office of the Department of the Interior. Maps may be purchased over the counter from the California State Office located on the 14th floor at 801 K Street, Sacramento, California. Folded maps are $10 each, or you can purchase rolled maps in a cardboard tube for $12. Rolled maps in their own storage tubes will last longer. Constant folding of maps deepens the creases until they split, resulting in tattered, useless maps. The tubes can be store behind the seat of your vehicle during field trips. More information on obtaining maps, satellite imagery, and aerial photographs may be found in the chapter, **"Outfitting the Prospector."**

Heading for the Hills!

It's time to go prospecting! You now know what you need to in order to head into the field and find gold for **fun** and for **profit**! The only thing left is to give you a hint about some gold-bearing locations where you can get you started. The list that follows is of gold-bearing areas where gold has been found both historically and in modern times.

You will need to do your prospecting to find out for yourself where the greatest concentrations of gold are located in any given area because high-concentration sites can vary from seasonally depending on rainfall and water movement. Be sure to respect all private property and valid claims.

The areas listed here are by no means the only ones where gold is located; they merely scratch the surface of the known gold-producing spots. There are many more gold-rich areas than can be crammed into one small chapter, but these will get you started. Remember to do your research!

Alaska

Alaska has so many gold-producing areas that an entire book would be needed to cover them all. However, a few to get you started would include the areas southeast and northeast of **Anchorage** around both Crow and Metal Creeks; also south of Anchorage about 50 miles in the Kenai Mountains. Southwest of Anchorage in the Cook Inlet area, check streams flowing into the Inlet. Valdez Creek north of Anchorage is a good prospect. Around the town of **Anvik,** check north along the Yukon river and the Flat Creek/Stuyahok River-area. Around **Bethel** gold has been found between the Kuskokwim and Stony River, the Arolik River south of Bethel, and the Kapon, Arsenics and Rainy Creeks-areas, and the creeks south of Bethel flowing into the Kuskokwin Bay. Seventy miles southeast of **Fairbanks** many of the north-flowing streams into the Tanana River contain gold; also southeast of Fairbanks to the Delta River. Check the areas north of Fairbanks. Prospectors have had luck along the Caribou, No Grub, Tenderfoot, Banner Creeks, as well as west of Fairbanks. Around Hot Springs, check tributaries flowing north into the Yukon River near Kallands and Fish Creek. Prospectors report gold along the Sullivan, Baker, and American Creeks. North of Hot Springs the Bear, Utopia, Red Mountain, Pocahontas, Indian, Snyder, and Black Creeks contain placer gold. Southwest of Fairbanks in Spruce Peak and the Busia Mountain are some prime placering areas. Around **Juneau** both lode and placer prospects are found near Brady Glacier, Windfall, Montana, McGinnis, Lemon and Nugget, Porcupine Creek, and also the Dundas River. North of **Livengood** check creeks flowing into Minook Creek; also check Quail, Gunnison, Troublesome, and Willow Creeks. Within about 50 miles of the town of **McGrath,** lode and placer gold is found in the Moore and Candle Creek areas. East of Prince William Sound check the Martin River, the White River, and the Yakataga Beach. South of the town of **Ruby** prospect Poorman, Long, Big Glacier, Ruby, and Baker Creeks.

Arizona

Arizona has many famous gold-producing areas. A few are covered here out of the many available. Four miles southeast of the town of **Bisbee** in Gold Gulch, and much of the Mule Mountains area north of Mexico, are good to fair producers. About six miles west of

Continental are the Amargosa Arroyo-area placers where flour gold is found. The **Dos Cabezas Mountains** are known for their placer deposits. **Dripping Springs Mountains,** at the south end, has some lode possibilities, while the southwestern side of Dripping Spring Wash has produced placer gold. Pinal Creek, upstream from **Globe,** is known for its placers. Northwest of Globe check out Gold and Lost Gulches, and Pinto Creek. Twelve miles southwest of **Hereford** placer gold is found. Approximately 60 miles north of **Kingman,** in the eastern part of the White Hills, placer gold is found. Three miles southwest of Kingman pocket gold has been found. Southeast of Kingman, in the northern reaches of the Hualapai Mountains, pocket and "wire" gold have been found. **Old Woman Gulch,** and other areas in San Domingo Wash, are good bets for placer miners. Mowry Wash and its tributary washes, located nine miles south of **Patagonia,** have been reported to contain rich placer gravel. Near **Payson** check out the Ox Bow Hill placers. Near the town of **Pearce** lode and placer gold is found in the vicinity of Pearce Hill. Southeast of **Prescott,** in the Bradshaw Mountains, is a good bet for placering with a metal detector. The eastern slopes of the Dome Rock Mountains between **Quartzite** and Blythe have produced some pocket and crystalline gold. East of Quartzsite, also in the Dome Rock Mountains, placer gold has been recovered. South of Quartzite, in the Plomosa Mountains placers are also located. Placer gold is reported on the eastern end of the Bradshaw Mountains. Eighteen miles south of **Topock,** in the Mojave Mountains, placers are located. Thirty miles north of **Tucson,** along the northwestern side of the Santa Catalina Mountains, placers have been worked over the years. **Willow Beach,** along the Colorado River, is known for course placer gold. **Wright Creek** and its tributaries has placer possibilities. Most rivers and creeks in **Yavapai County** are rife with recreational prospectors. There are a lot of claims in the area so be careful where you prospect.

California

One cannot think of California and not think of gold! There are hundreds of sites where recreational prospectors can recover both placer and lode gold in both the northern and southern parts of the state. The East Fork of the San Gabriel River in **Azusa Canyon** is a popular southern California gold prospecting area. The west slopes of the **Berkeley Hills** in Alameda County produced stringers of gold in quartz. One half mile south of **Big Bend Mountain,** in Pinkstown Ledge, gold in barite has been found. Approximately five miles west of **Big Pine,** on the southern flanks of the White Mountains, lode gold stringers are found in narrow quartz veins. Also, in the White Mountains area, some hefty gold nuggets have been recovered with metal detectors. **Butte Creek,** south of Chico to Centerville, is a good bet for the dredge, sluice, or gold pan. The **Feather River** around Oroville was worked extensively with bucket-line dredges, and the huge piles of rocks left by these dredging operations yield nuggets to those with metal detectors. Between the towns of **Lancaster** and **Mohave,** along State Highway 14, placer and lode gold is found. There are many valid claims in the area. **Piru Creek,** in the vicinity of Gold Hill Campground, has yielded placer gold. This area is currently closed for regrowth of vegetation destroyed by OHV (off-highway vehicles) use. Check with the Forest Service to see if the closure is still in effect. Near the town of **Randsburg,** in the Rand Mountains-area, are extensive dry placers. The Sacramento River and its tributaries, Oregon, Flat, and Clear Creeks, are good bets near the city of **Redding. Rich Bar,** a placer claim opened to public panning by Norm and Mike Grant, is a great place to hone your panning skills and recover some dandy nuggets. You may dig your own placer gravel, or purchase buckets for panning (no other equipment is allowed). For more information and current fees for **Rich Bar** prospecting, phone (916) 283-1730, or write to the Grants, c/o Rich Bar Mine, P.O. Box 39, Twain, CA, 95984. **Sierra County,** between Sierra City and Downieville, along the North Fork of the Yuba River, has yielded gold nuggets and lode gold to recreational prospectors. **Tehama County,** in the vicinity of Red Bluff and Jelly Ferry, has yielded placer gold to local prospectors. This area is relatively unknown to outsiders. Again be aware of many valid claims. North of **Twentynine Palms,** off State Highway 62, many of the desert washes are rife with placer gold. Much of it is too deep for metal detecting, but not all. Be sure to stay out of Joshua Tree National Monument when prospecting. Lode gold can be picked out of the mountains around Twentynine Palms.

Colorado

This is another state with hundreds of possible prospecting areas. Be aware that much of the gold production of this state was a by-product of the mining of other minerals. A few major gold producers are listed here. Much of the Colorado gold production is from lode mines. The upper banks of the **Blue River** have produced both lode and placer gold. **Chaffee County** has produced many ounces of placer gold, with the Arkansas River, Pine, and Clear Creeks being the most likely. Gold Run, Oregon, Lost Canyon, and Gilson Gulches are also placer prospects. In the area of **Gilpin**, placer possibilities abound. Near the town of **Georgetown**, placer and lode possibilities exist for the recreational gold prospector. The western flank of **Mount Wilson** has produced lode gold. The **Ophir Valley** is rich in lode deposits. Two miles north of **Pitkin**, both lode and placer gold are found. **Sacramento** and Beaver Creeks still yield gold for recreational prospectors. **Summit County** is another area famous for its gold production, as is **Teller County**.

Georgia

Georgia was the major U.S. gold producer until gold was discovered in California in 1848. Near the town of **Dahlonega** placer mining is quite popular. Expect to pay a small fee to prospect here. The **Etowah River** area is the site of several lodes. Creeks and streams in the **White County** area have yielded some placer gold.

Idaho

Idaho has numerous gold-bearing areas that are popular with recreational gold prospectors. Fifteen miles southeast of Boise, near **Arrowrock Dam**, is a spot known for placer gold production. **Blaine County** contains lodes with both gold and silver. **Boise County** is an area of considerable gold placers, some of which are still active. The South Fork of the **Boise River** is popular with recreational prospectors. The **Carrietown** area has both placer and lode possibilities. The **Caribou Mountains**, particularly around Mt. Pisgah, have produced both placer and lode gold. **Clearwater Creek,** near of the town of Pierce, has long been a placer gold area. The creeks of **Elmore County** have produced much placer gold. The area surrounding the town of **Gibbonsville** is known for both placer and lode gold. Approximately 40 miles south of **Grangeville**, in the French Creek area, good placer possibilities abound. Creeks and tributaries of **Idaho County** are known placer producers, as is the area around Elk City. Lode gold is reported in this area as well. **Jordan Creek**, in Custer County, has been a major placer producer. **Loon Creek**, in the vicinity of the ghost town of Casto, is a famous old placer area. **Napias Creek**, near the ghost town of Leesburg and west of the Salmon River, is a spot where rich placer deposits are located and still worked by recreational prospectors. There was some lode gold mined in there also. **Newsome Creek**, west of Elk City, is of interest to recreational prospectors. The area around the town of **Pioneerville** holds some lode possibilities. The areas around the town of **Quartzburg** are good spots to prospect for lode gold. Along the **Salmon River**, approximately 50 miles south of Grangeville, is a well-known placer area near the town of Riggons. The **Snake River** has been the site of placer mining since the late 1800s. The most likely spots are along the Snake River in Power, Bingham, Cassia, and Owhee Counties. Around the town of **Wallace**, in the Coeur d'Alene Mountains, lode gold is found. There are also placers which are worked on weekends by recreational gold prospectors. Gold-bearing lode veins may be located around the town of **Yellow Jacket**.

Maryland

Although gold is scarce here, there was some production in the area around **Great Falls**, east of the Potomac River. Lode gold has been found here also.

Michigan

Not much gold here either, but west of the town of **Ishpeming** are known quartz lodes.

Montana

The state of Montana has hundreds of gold-producing areas, too many to list. Here are a few of the popular ones. Twelve miles south of **Anaconda**, gold placers are known to exist. Ten to 12 miles northwest of Anaconda, lode gold is found. The vicinity of **Beaver Creek**, which empties into the Missouri River, has been a source of lode gold. The west slopes of the **Belt Mountains**, and many of the tributaries emptying into the Missouri River, are known placer producers. The **Blackfoot River**, and nearby Lincoln Gulch, have been major placer producers. East of **Boulder**, in the Elkhorn Mountains, lode gold is found. The southern slopes of **Bull Mountain** have produced veins with both silver and gold. **Confederate Gulch,** which empties into the Missouri River, is a known placer producer. At **Crow Creek,** and Johnny Gulch, near the town of Radersburg, lode and placer gold have been found. Lode gold is found approximately 12 miles west of the town of **Dillon**. The eastern slopes of the **Elkhorn Mountains** produced both placer and lode gold. In **Granite County,** near Bear Creek and Clark Fork River, placer gold is found. **Grasshopper Creek** in Beaverhead County is a site of placer gold. Some lode gold has been reported from this area. Ten miles south of **Helena,** placer gold is recovered. **Last Chance Gulch,** near Helena, is a famous placer area. Placers are also found northwest of Helena. As far as 12 miles west of the town of **Melrose,** lodes bearing both silver and gold are found. Ten miles north of the town of **Philipsburg,** along Henderson Creek, placer gold is found. Creeks and tributaries in **southwestern Montana** carry placer gold. The lode mines near the ghost town of **Sylvanite** offer some possibilities to prospectors with a metal detector. **Virginia Creek,** near Stemple Pass, contains placer gold.

Nevada

The slopes of the **Battle Mountains,** in the vicinity of the town of Battle Mountain, are the site of extensive placer fields. Approximately 40 miles south of **Boulder City,** near the old mining town of Searchlight, are known gold placers. **Buckeye Creek,** in the Pine Nut Mountains southeast of Reno, has yielded considerable placer gold. The **Bull Run Mountains,** specifically the along the Columbia, Sheridan, and Blue Jacket Creeks, have yielded much placer gold. In an area 25 miles northwest of **Carlin,** and around Maggie, Rodeo, Sheep, and Lynn Creeks, placer gold is found. Gold and silver lodes may be found on the western flanks of the **Clan Alpine Mountains** between Fallon and Austin. Many areas in **Douglas County** yield productive gold placers. **El Dorado Canyon** is the site of several small gold placers. Approximately 25 miles southeast of **Fallon,** in the vicinity of the Summit Mine drainage, placer gold has been recovered, and 40 miles southeast of Fallon, on Fairview Peak, lode and placer gold may be found. Also, south of Fallon, in both the Barnett Hills and Terrill Mountains gold has been located. Again, 40 miles southeast of Fallon, in Rawhide Wash, and in and around the Holligan and Balloon Hills, more placer fields are located. Areas around the town of **Genoa** are known for placer gold. **Gold Creek**, particularly in Hammond and Coleman Canyons, yielded an abundance of placer gold. Placer gold has been found in the areas surrounding **Goldfield**, and to the south, for approximately 12 miles. The area between **Gold Mountain** and Slate Ridge is reported to have several rich gold placers. **Grasshopper Gulch**, approximately 125 miles north of Elko, and surrounding areas, have yielded placer gold. The area around the town of **Jarbidge** has been the site of recreational prospecting for both lode gold and flour placers. Near **Johnnie**, southeast of Mercury, placer gold may be found both north and south of the town. **Lida Valley**, at the southern end, has been the site of small-scale placer mining. **Midas**, northwest of Elko, yielded both placer and lode gold. The **Monte Cristo Mountains**, particularly the northern side, have yielded placer and lode gold. Placer and lode mining have been conducted in the **Opal Mountains** in an area approximately 25 miles south of Boulder City. Seventy miles northeast of **Reno** placer gold has been found in the Trinity Mountains. Thirty miles east of Reno, recreational placering takes place in the Olinghouse Canyon and around Green Hill. In the **Santa Rosa Mountains**, in an area approximately 20 miles southeast of McDermitt, are outcroppings rich in gold ore. Great "pocket gold" possibilities exist here. In the same vicinity, placer gold has been recovered along Rebel, Willow, Canyon, and Pole Creeks. **Sawtooth Knob**, approximately 65 miles northwest of Lovelock, is the site of several placering areas. The **South Fork** of the Owyhee River has been prospected for placer gold. Lode gold also has been found nearby. Approximately 60 miles southwest of Tonopah is **Tule Canyon**, located between the Magruder and Sylvania Mountains. Placer mining has been done here. Also in this area, you might try placering near the eastern slopes of the Sylvania Mountains, as well as Pigeon Spring and Leadville Canyon. **Temple Bar**, in Lake Mead, yields placer gold to dredgers, as do several nearby areas. **Trinity Canyon**, 10 miles due north of Lovelock, has been the site of major placer activity. **Van Duzer Creek** is reported to contain placer gold. Twenty miles northeast of **Winnemucca** is a placering area.

New Mexico

Approximately 25 miles northeast of **Albuquerque**, placer gold has been recovered from Tejon Canyon and Las Huertas Creek. The eastern end of the **Animas Hills**, particularly around the gulches, Gold Run, Green Horn, Hunkydory, Dutch, Grayback, and Little Gold Run, were the richest placer areas. **Bald Mountain**, in the Elizabeth Town-area is a good placer prospect, especially around Willow, Ute, and South Pontil Creeks. **Bear Creek**, and the nearby Santo Domingo Gulch, around the town of Pinos Altos, are rich placer areas. The southern slopes of the **Big Burro Mountains** have been the site of extensive placering. On the western slopes of the **Caballo Mountains**, near the reservoir, placer gold is found. In the vicinity of the town of **Dolores**, in Cunningham Gulch, placer gold has been located. Near the town of **Golden**, on the west slopes of the San Pedro Mountains, placer gold has been found. The best sites were Lazarus Gulch and Tuerto Creek, as well as the washes and tributaries leading into them. The **Jicarilla Mountains**, near the town of the same name, have been the site of much placer mining. The Ancho, Spring, Warner, and Rico Gulches were noted as the richest placer sites. Placer gold is found in many of the gulches draining into the **Moreno River**, which is between 30 and 40 miles northeast of Taos. The two richest gulches are the Mills and Anniseta. In the **Ortiz Mountains**, a few miles north of the towns of Cerrillos and Golden, placer gold has been found. In and around **Placer Creek**, near the town of Red River, placer gold is found. Near the confluence of the Red River and the Rio Grande, and approximately 30 miles south to the country line, placer gold is recovered. Other placers in the area are found near Lama, Alamo, Garrapta Canyons and Cabresto Creek. Approximately 10 miles east of **Silver City** are several major placer areas.

North Carolina

Northwest of Asheboro, along the Uwharrie River, placer and lode gold have been found. Approximately two miles west of **Candor**, lode gold has been found. Southeast of **Concord**, lode gold has been found. Three miles south of **Georgeville**, nuggets have been recovered from streams. The gravel of **Georgetown Creek**, just south of the Transylvania County line, has yielded placer gold. South of **King's Mountain**, about two miles, gold has been found. A small amount of lode gold has also been reported. Near the town of **New London**, in the Parker Mine area, lode and "pocket gold" have been found. A couple of miles west of the town of **Ophir**, placer gold was found along the Uwharrie River. South of Morganton, in the **South Mountains**, placer gold has been recovered. Lode and placer gold have been found approximately three miles northwest of Waxhaw.

Oregon

Near the town of **Applegate**, and along the river of the same name, placer gold has been found. Areas around **Auburn**, in Baker County, have yielded placer gold. Northeast of **Baker**, the Griffin Gulch has yielded placer gold. Also, both southwest, and 10 miles northeast, of Baker, placer and lode gold have been found. West of Baker, in Grant County, are extensive placers. Lode mines near **Bourne** have produced gold. Approximately 10 miles southeast of **Bridgeport**, and between Bridgeport and Hereford, placer gold has been found. Lode gold has been found around the town of **Cornucopia**. Southwest of Cornucopia, both Eagle and Paddy Creeks, in the Wallowa Mountains, have yielded both lode and placer gold. Also, the area south of Cornucopia has yielded placer gold. **Desolation Creek** has yielded placer gold. **Elk Creek**, near Susanville, has produced much placer gold. Areas around the town of Galice have produced lode and placer gold. **Gold Hill** is surrounded by placer areas. The area around the town of **Greenhorn** has yielded lode and placer gold. **Jackson Creek** has been the site of much placer gold recovery. Downstream of the confluence of **Josephine Creek** and the Illinois River, placer gold has been recovered. The North Fork of the **John Day River**, near the north end of Grant County, has yielded placer gold, as has the Middle Fork of the same river.

South Carolina

Two miles south of the town of **Jefferson**, lode and placer gold have been found. Three miles northeast of the town of **Kershaw**, near the Haile Mine, lode gold is found.

South Dakota

The area around **Deadwood** has produced both lode and placer gold. Four miles south of **Lead**, lode gold has been found. The area surrounding **Hill City** has produced lode gold. Also, areas northwest of Lead have produced lode gold. **Ragged Top**, northwest of Lead has produced gold.

Utah

The southern slopes of the **Abajo Mountains**, particularly along Recapture and Johnson Creeks, have yielded flour gold. **Amasa Valley**, 45 miles southwest of Delta, has been reported to have scattered placer gold deposits. Miller Canyon, in the same general area, contains placer gold. The **Colorado River**, near Glen Canyon, has placer gold. The **Deep Creek Mountains** have scattered placer gold deposits. The **Green River** is reported to have scattered placers of flour gold, particularly near the town of Vernal. Near the town of **Jensen**, in the area of Horseshoe Bend, below the Split Mountain Gorge, placer gold has been found. The area south of **Marysvale**, in and around the Pine Gulch Creek, has placer gold. Twenty-five miles northeast of **Moab**, in Miner's Basin, Placer Creek and the Wilson Mesa area, placer gold has been found. The eastern slope of **Mount Ellen**, in the vicinity of Crescent Creek, has some placer gold. The **Wasatch Mountains**, approximately 20 miles southeast of Salt Lake City, have produced placer gold.

Virginia

Busby Creek produced placer gold, and lode gold was found in the area of the Goochland and Fluvannan County lines. Approximately 18 miles west of **Fredricksburg**, lode gold was found. Approximately three miles south of **Morrisville**, placer and lode gold have been found. Placer and lode gold are also found along the **Rappahannock River**.

Washington

The eastern slopes of the **Cascade Range** have produced placer and lode gold. Near the town of **Chewsaw**, along Mary Ann Creek, placer and lode gold have been found. The **Entiat River** is the site of several gold placers. Near the headwaters of **Granite Creek**, placer gold has been recovered. **Palmer Mountain**, in Okanogan County, is the site of numerous rich placers. **Peshastin Creek**, near the town of Wenatchee, has been the site of several productive placers. Placer gold has been located along the Columbia River near the town of **Republic**. **Squaw Creek**, approximately nine miles north of the Methow River, is a lode and placer gold area. **Ruby Creek**, in Whatcom County, has produced some placer gold.

Wyoming

Douglas Creek, in Albany County, has yielded some placer gold. Lode gold comes from the same area. Placer and lode gold have been found 25 miles south of the town of **Lander**.

Western Australia

Western Australia has numerous sites for locating gold. A few are listed here. The small town of **Cue** has been the site of many spectacular nugget finds. The **Eastern Gold Fields** of Western Australia are popular with prospectors wielding metal detectors. **Laverton** is a town surrounded by placer fields. **Lawlers** is another great placer area. Areas around the town of **Leonora**, in almost any direction, are rich in placer gold. **Meekatharra** is another popular spot. Near Leonora is **Mertondale**, which has yielded sizable nuggets to prospectors with metal detectors. The **Murchison Peak** area is another popular spot. **Nullagine** is surrounded by placer fields. **Peak Hill** yields placer gold. The town of **Wiluna** is surrounded by gold placers.

Outfitting The Prospector

In this chapter you will find mail-order sources and maps for all your gold prospecting needs. I have listed a few sources here, but there are so many that it is not feasible to list all of them. The best idea is to check the yellow pages of your phone directory under the following classifications:

Gold Prospecting Supplies

Mining Equipment

Mail-Order Sources

Geological Supply Catalog, Contract Geological Services, Inc., (800) 247-6853, (702) 358-0923, FAX (702) 358-8209. Rock picks, hoe picks, five-gallon plastic buckets, screens, sample bags, gold pans. A wide variety of supplies for all geological field work. No gold recovery equipment.

Fisher Research Laboratory, (209) 826-3292, FAX (209) 826-0416, (800) 672-6731 to mail-order. Metal detectors & detecting equipment, books, videos. Call for a list of dealers near you. One of the best!

Cal-Gold, (626) 792-6161. Gold recovery equipment, gold pans, retorts, amalgamation equipment, mining supplies, books, videos, maps, assay equipment, picks, magnifying loupes. One of the best!

Keene Engineering, (818) 993-0411, FAX (818) 933-0447. <mobilmark@aol.com. Recovery equipment, gold pans, picks. Call for a list of dealers near you. One of the best!

Kellyco, (800) 327-9697, (407) 699-8700, (407) 695-7700. Recovery equipment, treasure hunting supplies, videos, collectible firearms, gold pans, books, maps. Discounted equipment!

Pedersen's, (714) 771-6463 for information, (800) 953-3832 for mail-order. Prospecting equipment.

Arizona Al's, (602) 930-1755 for information, (800) 987-7795 for mail-order. <Arizona.Al@postoffice.worldnet.att.net>

On-Line Catalogs

Cal-Gold, (626) 792-6161. <http://www.treasure.com/calgold/freecat.htm> for a free catalog. <http://www.treasure.com/calgold/prospect.htm> for an equipment price list, or type "Cal-Gold" for Internet search. Gold recovery equipment, gold pans, retorts, amalgamation equipment, mining supplies, books, videos, maps, assay equipment, picks, magnifying loupes. One of the best!

Northwest Treasure Supply, (425) 641-1325, FAX (425) 881-7340, (800) 845-5258. <http://www.treasurenet.com/nwts> Picks, sluice boxes, gold pans, detectors, books, videos.

Bullion Bob's, bbob@halcyon.com. Recovery equipment, gold pans, picks.

Maps

Big Ten, Inc. Gold Maps, P.O. Box 321231-W, Cocoa Beach, FL 32932-1231, (407) 783-4595, < http://www.megabits.net/gold/>, or just type "Big Ten Maps" for Internet search. Maps show hundreds of locations where gold has been found historically and where prospectors are finding gold today! Individual and map sets for California, Alabama, Georgia, North Carolina, South Carolina, Virginia. Check on status of map publication for other states.

U.S. Geological Survey (USGS) Topographic Maps, Map Distribution Section, U.S. Geological Survey, Box 25286, Federal Center, Denver, CO 80225. Direct mailing source for topographic maps for all states in the U.S. Request the topographic map index. It's free!

Then you can mail-order the desired maps. Both 15-minute and 7.5-minute maps are available. Shows elevations, mines, all roads, intermittent water flow, etc. Range and Township numbers. Allow at least six to eight weeks for delivery.

DeLorme Atlases, DeLorme Mapping Company, P.O. Box 298-5400, Freeport, Maine 04032. Atlases for Maine, New Hampshire, Vermont, Northern California, Southern & Central California, Florida, Michigan, Nevada, New Mexico, New York, Ohio, Pennsylvania, Tennessee, Utah, Virginia, Washington, Wisconsin, Minnesota, Illinois. Topographic maps show most features, but no Range and Township numbers.

A prospecting convention like this one is a great place to shop for equipment.

California Geologic Maps, California Division of Mines and Geology, P.O. Box 2980, Sacramento, CA 95812-2980, (916) 445-5716. Depicts rock-type by color. For other states, contact the state office or land management officer in that state for information.

Satellite Imagery Photographs, USGS Eros Data Center, Sioux Falls, South Dakota 57198 (605) 594-6511.

ESIC-W/U.S. Geological Survey, 345 Middlefield Road, MS 532, Menlo Park, CA 94025 (650) 329-4390. **Aerial photographs** may be ordered as 9x9 semi-glossy prints, subject to availability.

U.S. Forest Service, Regional Office, 630 Sansome Street, San Francisco, CA 94111 (415) 705-2874. Inquire as to price and availability.

Surface and Mineral Ownership Maps, Bureau of Land Management, California State Office, 2135 Butano Drive, Sacramento, California 95825 (916) 979-2800.

Publications & Organizations To Assist You

In this chapter you will find information on magazines, books, clubs and other organizations that can be of help to the recreational gold prospector.

There are magazines that cover all aspects of recreational gold prospecting, including treasure hunting. In magazines you can get tips on building your own equipment or improving recovery techniques, read inspiring stories by prospectors who are finding gold, and get valuable suggestions on where to prospect. The classified sections of prospecting magazines are great places to find used equipment

Magazines offer up-to-date information on the ever-changing legislative status of recreational prospecting – a "must-know" for the prospector! A trip to your local newsstand will give a good indication of the many available publications.

Books are wonderful sources of information on all aspects of recreational gold prospecting. Many offer information on geology – the knowledge of which can certainly enhance one's ability to locate gold...especially lode gold. Some of these books are written for novices, others for professional miners. All have a wealth of information that the prospector can use in his or her search for gold.

Clubs and organizations with a gold prospecting focus can put you into contact with other people who enjoy recreational gold prospecting. There are many benefits to club membership. Many prospecting clubs own their own gold claims. Members also benefit from activities and trips centered around the fun of gold prospecting. Much can be learned from the experiences of club members who have been longtime prospectors.

Magazines and Newsletters

International California Mining Journal, P.O. Box 2260, 9011 Soquel Drive, Aptos, CA 95001. (408) 662-2899. A good read for the serious prospector who is also interested in commercial mining and mining legislation. The classified section of this publication is full of ads for used equipment!

Rock & Gem Magazine, 4880 Market Street, Ventura, CA 93003. (805) 644-3824. A magazine devoted to outdoor enthusiasts, mineral collectors and lapidaries. The January and July issues are devoted solely to articles on recreational gold prospecting.

Fisher World Treasure News, Dept. NL-52, 200 Willmott Road, Los Banos, CA 93635. (209) 826-3292. A great source of information for the electronic prospecting enthusiast and treasure hunter. Covers all aspects of metal detecting and "how-to" methods.

Lost Treasure, Inc., P.O. Box 451589, Grove, OK 74345. (918) 786-2182. A great one for recreational prospectors and treasure hunters. Lots of good stuff on electronic prospecting.

Treasure Facts, a publication of **Lost Treasure, Inc.,** great resource for "how-to" articles on techniques that will increase your prospecting and treasure hunting success. Also has an e-mail address, <http://www.losttreasure.com>. Classifieds feature used equipment.

Gold and Treasure Hunter, P.O. Box 47, Happy Camp, CA 96039. A good one for recreational gold prospectors.

Pick & Shovel Gazette, P.O. Box 507, Bonsall, CA 92003.

Treasure, P.O. Box 489, Yucaipa, CA 92399.

Western & Eastern Treasures, P.O. Box 1598, Mercer Island, WA 98040-1598. Gold prospecting and treasure hunting. Classifieds feature used equipment.

Treasure Hunter Confidential, P.O. Box 8162-H, Rancho Santa Fe, CA 93067

Popular Mining, Action Mining Plaza, 4469 W. Reno Avenue, Las Vegas, NV 89118. A great one for the do-it-yourselfer!

Gold Prospector, P.O. Box 507, Bonsall, CA 92003. A good one for the recreational gold prospector.

Books

Handbook For Prospectors, **Fifth Edition**, by Richard M. Pearl, McGraw-Hill Book Company, New York, NY: 1973. One of the all-time best! An update of M. W. von Bernewitz classic, "Handbook for Prospectors and Operators of Small Mines."

Handbook For Prospectors and Operators of Small Mines, by M. W. von Bernewitz, McGraw-Hill, New York, NY: 1926. This book is out-of-print, but may be found in used book stores with some diligent hunting.

Operating Small Gold Placers, by William F. Boericke, John Wiley & Sons, Inc., New York, NY: 1936. Another out-of-print book worth searching for. Useful information for the do-it-yourselfer.

Where To Find Gold In ... series of books, by James Klein, Gem Guides Book Company, Baldwin Park, CA: 1994. Good basic information where and how to find gold in California.

Gold Diggers Atlas, by Robert Neil Johnson, Cy Johnson & Sons Book Distributors, Susanville, CA: 1975. A guide to finding gold locations in the Western United States.

Methods of Placer Mining, Basque, Sunfire Publications, Langley, B.C., Canada, 1994. Discusses the various gold rush areas of North America and explains recovery methods.

Diving And Digging For Gold, by Mary Hill, Naturegraph Publishers, Happy Camp, CA: 1974. A small book packed with useful information on dredging for gold and gold amalgamation techniques. Some gold locations.

Modern Metal Detectors, Garrett, RAM Publications, Dallas, TX: 1995. This book contains complete information about all conventional types of detector, and has been newly revised to include computerized detectors as well.

The Weekend Gold Miner, by A. H. Ryan, Ph.D., Gem Guides Book Company, Baldwin Park, CA: 1978. A small book packed with useful information on electronic prospecting, desert prospecting, how-to build a retort, and some gold locations.

Gold Fever And The Art Of Panning And Sluicing, by Lois DeLorenzo, Gem Guides Book Company, Baldwin Park, CA: 1978. One of the best on basic recreational gold prospecting. Good plans and diagrams for building your own sluice box. Includes some gold locations.

Advanced Nugget Hunting, Pieter Heyerdelaar & David Johnson, Fisher Research Laboratory, Los Banos, CA: 1988. Great how-to on electronic gold prospecting.

Treasure At Your Feet, by Paul Heuser, Heritage Sales, Stoughton, WI: 1984. A good basic introduction to treasure hunting with a metal detector.

Successful Nugget Hunting, Volume I, by Pieter Heydelaar, Fisher Research Laboratory, Los Banos, CA: 1991. How-to on using a metal detector to find gold. Great information and maps on where to prospect in Australia, California and Arizona.

Suggestions For Prospecting, a publication of the U.S. Geological Survey, Denver, CO: 1989. Covers a variety of sampling techniques, such as geophysical, biological, magnetic, radiometric. Good reference for books on delving further into these subjects, also map ordering information.

Techniques of Pendulum Dowsing, by Bill Cox, Forces, Santa Monica, CA: 1977. Includes map dowsing.

Techniques of Swing-Rod Dowsing, by Bill Cox, Forces, Santa Monica, CA: 1977. Techniques for using dowsing rods.

Prospecting For Gemstones And Minerals, by John Sinkankas, Van Nostrand Reinhold Company, New York, NY: 1970. Contains useful information for the prospector interested in geology associated with all aspects of prospecting.

Location and Patenting of Mining Claims and Mill Sites In California, a Bureau of Land Management publication: 1996. Contains useful, straightforward information on locating and claiming "locatable" minerals. The book also includes the various forms that are required for filing on various aspects of concerning your claims. The book can be ordered by phone for $6 and charged to your credit card. A nice advantage to ordering maps or books from the BLM is that they ship your order immediately. I usually receive an order within just a couple of days of placing my order. Too, the BLM employees have always been friendly and helpful.

Contemporary Wire Wrapped Jewelry, by Curtis Kenneth Leonard with William A. Kappele. Ideas and instructions for using gold wire for creating wearable jewelry. The ideas in this book can be adapted for creating gold nugget jewelry.

Organizations & Clubs

New clubs are forming all the time. If you cannot locate one in your area, you might consider starting your own! All you need are a few like-minded souls, and an interest in gold prospecting!

ALASKA

Alaska Treasure Seekers Society
200 W. 34th #825
ANCHORAGE, ALASKA 99503
E-mail: <aleigh@ak.anc.net>

ARIZONA

Gold Prospectors Association
of Tucson (GPAT)
5425 E. Broadway #216
Tucson, AZ 85711
E-mail: <gpat@flash.net>

Havasu Gold Seekers Club
P.O. Box 3281
Lake Havasu City, AZ 86405-3281

Mesa Gold Diggers
7520 Paseo Escondido
Prescott Valley, AZ 86314
(520) 772-8922

Mohave Prospectors Association
P.O. Box 6446
Kingman, AZ 86402-6446

Roadrunner Prospectors' Club
P.O. Box 56804
Phoenix, AZ 85079
Office: (602) 274-2521
Fax: (602) 274-4335

CALIFORNIA

Antelope Valley Treasure
Hunters Society
P.O. Box 4708
Lancaster, CA 93534

Coarsegold Gold
Prospectors Club
35463 Highway 41
Coarsegold CA 93614

Gold Country Miners
P.O. Box 899
Foresthill, CA 95631

Gold Nugget Prospectors Club
9430 Mission
Riverside, CA 92509
(909) 685-5531

Golden Caribou Mining Club
Belden, CA (Feather River area)
Steve Draper
(916)283-0956
E-mail: caribou1@psln.com

High Desert Gold Diggers
20162 Highway 18, Suite G-178
Apple Valley CA 92307

Mohave Prospectors Association
P.O. Box 6446
Kingman, AZ 86402-6446

Motherlode Goldhounds
P.O. Box 4389
Auburn, CA 95604

Orange County 49ers, Inc.
Pete Siedzick
201 N. Wayfield St., #48
Orange, CA 92667
(714)289-8027
E-mail: <pete@midcom.com>

Sacramento Detecting Buffs
4636 Nottingham Circle
Sacramento, CA 95864

Santa Clara Miners Assoc.
P.O. Box 2861
Sunnyvale, CA 94087-2861

The Prospectors Club
    of Southern California
5825 Rowland Ave.
Temple City, CA 91780

The Santa Rosa Gold Diggers
Jim Zambenini
1707 Mission Blvd.
Santa Rosa CA 5409
(707) 538-2984
E-mail: <zam@netdex.com>

The Southwestern Prospectors
    and Miners Association
P.O. Box 904
La Mesa, CA 91944-0904
(619) 486-9161

United Prospectors Inc.
7425 English Hills Rd.
Vacaville, CA 95688-9528

Valley Prospectors Club
Dee Stapp
(909) 883-7303

West End Prospectors Club
P.O. Box 834
Fontana, CA 92335

West Bay Prospectors & Metal
    Detectors (Nice site)
San Carlos, CA
Alan Chun (415) 349-0937
<Alanch@hooked.net>

COLORADO

Chaffee County Prospectors Club
P.O. Box 3051
Buena Vista, CO 81211
E-mail: <mlallen@rmi.net>

Gold Prospectors of Colorado
P.O. Box 1593
Colorado Springs, CO 80901-1593
Gary S. Turk President
(719) 390-7355
E-mail: <daddio-
    1@ix.netcom.com>

Rocky Mountain Prospectors
    and Treasure Hunters Club
P.O. Box 1483
LaPorte, CO. 80535
(970) 221-1062

The Gold Prospectors of the
    Rockies
7811 West Mississippi
Lakewood, CO 80226
(303) 980-8430
Fax (303)274-1593
E-mail: <stevej@ix.netcom.com>

GEORGIA

Weekend Gold Miners
Glenn R. Davis
P.O. Box 7142
North Georgia College
Dahlonega GA, 30597

IDAHO

Idaho Gold Prospector Associa-
    tion
5027 Cree Way
Boise, ID 83709

Northwest Gold Prospectors
P.O. Box 1736
Hayden Lake, ID 83835
Phone/Fax (208)687-2382
E-mail: Bob Lowe
    <rlowe@rand.nidlink.com>

MASSACHUSETTS

Gold Miners of New England
P.O. Box 525
Athol, MA. 01331

NEVADA

Comstock Gold Prospectors
P.O. Box 20781
Reno, NV 89515

Gold Searchers of Southern
Nevada
P.O. Box 96732-6732
Las Vegas NV
(702) 393-GOLD (393-4653)

NEW HAMPSHIRE

Northeast Recreational Gold
Miners Association (NRGMA)
P.O. Box 223
Jaffery, NH 03452

Recreational Detecting and
Prospecting Club
26 S. Main St.
Concord, NH 03301-4848
(603) 224-5909

NEW MEXICO

Gold Prospectors Association of
New Mexico
P.O. Box 25472
Albuquerque, NM 87125

New Mexico Gold Miners
Association
3900 Delamar
Albuquerque, NM 87110

UTAH

Northern Utah Prospector's
Association
Ogden, UT
Curt Dayton (801) 392-3257
Alan Meyer
<alan.meyer@m.k12.ut.us>

Utah Gold Prospectors Club
Paul Bidali, President
100 S. Fort Lane
Layton, UT 84041
(801) 546-4086

WASHINGTON

Northwest Mineral Prospectors
P.O. Box 1973
Vancouver, WA 98686

Washington State Prospectors
  Mining Assn.
W.P.M.A.
10002 Aurora Avenue North
  #1193
Seattle, WA 98133
(206) 784-6039
E-mail: Chuck Cox <au-
  miner@accessone.com>

Washington Alliance of Miners
  and Prospectors
P.O. Box 1363
Monroe, WA 98272

Boeing Employee Prospectors
  Society
(Everett & Kent Organizations)
P.O. Box 3707
Seattle, WA 98124

No-Name Prospectors
P.O. Box 2872
Woodinville, WA 98072

Northwest Prospectors
20902-67th Ave. NE #343
Arlington, WA 98223

Yakima Prospectors
2006 S. 10th Ave.
Yakima, WA 98903

Bedrock Prospectors
7507 214th Ave. E.
Sumner, WA 98390

Northwest Gold & Gem
  Prospectors
2616 Bridgeport Way West
Tacoma, WA 98466

N. Central Washington
  Prospectors
914 Monroe St.
Wenatchee, WA 98801

Okanagen Prospecting and
  Mining Assoc.
P.O. Box 47
Riverside, WA 98849

Gold & Treasure Hunters Club
2009 Iron St.
Bellingham, WA 98225
Ethan Jones
<bigfoot@eskimo.com>

INTERNATIONAL

Australia

Brisbane Prospecting Club
Brisbane, Australia
Danny Van der Walle
E-mail: <danny@ats.com.au>

Geelong Prospecting Club
Sheila Middleton
P.O. Box 348
Belmont 3216
Victoria, Australia

Canada

Alberta Gold Prospectors
    Association
3020 Sanctuary Road
Calgary T2G 5C9
E-mail: Marj Lawrence
    <wyz@mail.planet.eon.net>

Vernon Placer Miner's Club
Box 1691
Vernon B.C. Canada
V1T 8C3
E-mail: <castrucw@junction.net>

France

Federation Francaise
    d'Orpaillage
Salle 605
Complexe de la Republique
64000 PAU

RHON'OR
c/o Jacques Brest
6 rue V. Komarov
69200 Venissieux
France
Tel: +33/4-72-51-00-69

Finland

Mr. Kauko Launonen
Goldmuseum
Goldvillage
99695 Tankavaara

Germany

German Goldpanning
    Association
KIrchgasse 25
95497 Goldkronach Germany

Sweden

Guld Graevarnas Foerening
    (or simply GGF)
Mr. Lars Guldstrom
S-570 15 Holsbybrunn
Sweden
Phone 383-40110, Fax 383-40109
E-mail:
    <goldwashing@guldstrom.se>

Ovre Norrlands Guldgravarklubb
c/o Greta Savela
Stormvagen 343
SE-97634 Lulea, Sweden

Svenska guldvaskarfoereningen
Johnny Hagberg, President
Moraenvaegen 26
S-13651 Haninge
Sweden

Switzerland

Schweizerische
Goldwdschervereinigung
Peter Pfander
Schwanden 32
3054 Schupfen
Switzerland

United Kingdom

UK Goldpanner's Association
c/o Mick Gossage
12 Pikepurse Lane
Richmond
Yorkshire DL10 4PS

Government Agencies
UNITED STATES DEPARTMENT OF THE INTERIOR, BUREAU OF LAND MANAGEMENT STATE OFFICES

Alaska State Office
222 W. 7th Ave. #13
Anchorage, AK 99513
(907) 271-5960

Arizona State Office
P.O. Box 555
Phoenix, AZ 85001 or,
222 North Central
Phoenix AZ 85004
(602) 417-9200

California State Office
2135 Butano Dr.
Sacramento, CA 95825
(916) 979-2800

Colorado State Office
2850 Youngfield St.
Lakewood, CO 80215
(303) 239-3600

Eastern States Office
7450 Boston Blvd.
Springfield, VA 22153
(703) 440-1600

Idaho State Office
1387 S. Vennell Way
Boise, ID 83709
(208) 373-3891

Montana State Office
(N. Dakota & S. Dakota)
Granite Tower
P.O. Box 36800 or,
222 N. 32nd St.
Billings, MT 59107
(406) 255-2885

Nevada State Office
850 Harvard Way
P.O. Box 12000
Reno NV 89502
(702) 785-6500/6505

New Mexico State Office
1474 Rodeo Road
P.O. Box 27115
Santa Fe, NM 87504-1449
(505) 438-7575

Oregon State Office
(Washington)
P.O. Box 2965
Portland, OR 97208
(503) 952-6001

Utah State Office
324 South State St., Suite 301
Salt Lake City, UT 84111-2303
(801) 539-4001

Wyoming State Office
2515 Warren Ave.
P.O. Box 1828
Cheyenne, WY 82003
(307) 775-6256

For Nebraska
New Castle RA
1101 Washington Blvd.
Newcastle, WY 82701-2972
(307) 746-4453

The Bureau of Land Management offers a number of publications and maps that may be of particular interest to the recreational gold prospector. Following is a list and price information:

Camping
California Camping $19.95
Easy Camping in California $12.95

Minerals
16 Specimen Mineral Kit & Study Guide $9.95
16 Specimen Rock Kit & Study Guide $9.95
Field Guide To Geology $14.95
Gem Trails of Southern California $8.95
Gold Prospector's Handbook $10.95
AAA Mother Lode Guide Map $3.95
Peterson's Guide – Rocks & Minerals $4.95
Rockhound and Prospector's Bible $9.95
Rockhound's Guide To California $12.95
The Weekend Gold Miner $5.50

BLM Maps
Northern California Atlas & Gazetteer $16.95
Southern California Atlas & Gazetteer $16.95
The following maps must be ordered by area or name:
Forest Service Maps $7.00
USGS Topographic Maps $5.00
Surface/Mineral Management Maps $4.00
Desert Access Guides $4.00
American Discovery Trail Map (California) $7.95
Ancestral Yuba River Gold Map $5.99

About the Author

Gail Butler learned to pan and process gold ore when she was ten years old by watching her prospector grandfather. She has been a devoted recreational gold prospector since a weekend trip to southern California's San Gabriel Mountains in 1979, when she panned a few flakes with his old copper pan and rock pick.

Gail retired from the Los Angeles County Sheriff's Department in 1993. Now she writes about gold prospecting for a variety of publications.

Gail is a contributing editor for *Rock and Gem* magazine. Her first book, *The Rockhound's Guide to California,* was published in 1995.